the UNCOMMON LIFE
of              OBJECTS
        COMMON

The UNCOMMON LIFE
of COMMON OBJECTS

ESSAYS on Design and the Everyday

Akiko Busch

METROPOLIS BOOKS

Editor: Diana Murphy
Managing editor: Melanie Archer
Designer: Meryl Pollen
Separations and printing: CS Graphics PTE Ltd., Singapore

This book is set in Garamond and printed on SR matte

Library of Congress Cataloging-in-Publication Data

Busch, Akiko.
    The UNCOMMON LIFE of COMMON OBJECTS / Akiko Busch.–1st ed.
    p. cm.
    ISBN 1-933045-06-X
    1. Busch, Akiko. 2. Design, Industrial. I. Title.
    TS171.4.B874 2005
    745.2–dc22
                            2004030307

Metropolis Books is a joint publishing program of:

D.A.P./Distributed Art Publishers, Inc.
155 Sixth Avenue, 2nd floor
New York NY 10013
tel 212 627 1999  fax 212 627 9484
www.artbook.com

and

*Metropolis* Magazine
61 West 23rd Street, 4th floor
New York NY 10010
tel 212 627 9977  fax 212 627 9988
www.metropolismag.com

Available through D.A.P./Distributed Art Publishers, Inc., New York

FOR my sister, MARY F. BUSCH

# CONTENTS

FOREWORD

When Akiko Busch first proposed writing *The Uncommon Life of Common Objects*, I went into a mode of anticipation. Even as she worked in her quiet, gentle, but deliberate way on the manuscript in her home office in rural upstate New York, I was imagining hours spent on weekend afternoons in my tiny Manhattan loft, curled up in my favorite chair, reading her views on life and design. That's what Aki does really well: she's able to bring human behavior and aspiration together with design activities and outcomes. And when she introduces her twin boys into the discussion, she also brings the next generation into the dialogue.

When you consider that everything around us is designed, more or less successfully, you understand the importance of designers to a world made of parks, streets, buildings, rooms, objects, and signs – myriad things that can ease or frustrate our activities, that can delight or madden us. Aki gets to the heart of our world of design when she essays the objects we use every day. As she introduces us to their complex stories, we understand, or even redefine, our relationship with our own possessions.

For designers, Aki's narrative is essential information. Knowing people's intimate and storied relationships with the things they own and use is something I wish all designers took seriously. If they did, the fit between ourselves and our objects would be much more comfortable. For the rest of us, Aki's words are an encouragement to value our possessions, not for their high prices or relative coolness, but for what they mean to us.

You are about to enter a very personal world with an acute observer of people and design as your guide. Read on and find yourself there.

susan s. szenasy

INTRODUCTION

Like most children, my sons went through a period when they were enchanted by magic tricks, and they prized their collection of two-headed nickels, knotted scarves, metal hoops that seemed to connect and disconnect at will, and assorted other props of deception. But among their favorites was a small, flat wooden box with the picture of a house on it. The box had a small slot, which contained a drawer with a thin disc, just big enough to hold a quarter, carved in its surface. Depending upon how the box was held and its drawer maneuvered, the quarter could appear, or disappear, giving the box its value or, in turn, taking it away. And for months, practicing this trick was a cherished exercise. Sometimes the little box had value, sometimes it had none, and its worth seemed to come and go, by pure chance. To the viewer, the selection appeared random and impossible to predict, but to the child, the logic of the trick, though secret, was clear, explicable, precise.

I couldn't help but think of this little wooden box with its two drawers as my boys grew older and started to prize their *things* – first their action figures and basketball cards, then later their snowboards, backpacks, cameras. And the questions kept coming up: What gives ordinary objects their value? Where is the quarter hidden this time? Who has put the coin in the drawer, and what is its value?

How we assign merit to things often seems motivated by the vagaries of human impulse; the process is improvisational and at times comic. As a writer, I have often been given objects for articles and essays I have written. Over the years, I have been given a brass umbrella stand, a small red clock, a blue canvas bag, a table-cloth, a crystal shot glass, a martini pitcher. I accepted them all. Notwithstanding the conventional journalistic ethos of not taking favors and gifts from the subjects of one's stories, I accepted these things because I am charmed and intrigued by the impulse of such exchanges. I don't know that these are even exchanges, but I don't know that in this particular realm there is such a thing as an even exchange. The list of items I have been given has no particular logic, which is probably as it should be; there isn't much logic in trying to match objects to words. Still, I put a high value on the effort *anyone* makes to try to make such matches. And so I think, maybe that's it: a tablecloth is worth five hundred words; an umbrella stand is worth a thousand words; a martini pitcher, two thousand. All of these seem as good a place to start as any.

I don't think I am the only one curious about how we assign value to things. What is known as the American collectibles market generates over $10 billion annually. Auction houses selling the clothes, the furniture, the vast archives of memorabilia that once belonged to such public figures as Jackie Onassis, the Princess of Wales, Marilyn Monroe, and Leonard Bernstein routinely clear double and triple the anticipated amounts. Who could have known, the auctioneers ask. But if other people's things hold out an allure, so do our own. Each week, Americans across the country exhibit what may – or may not – be their treasures on *Antiques Roadshow*, displaying everything from scrimshaw mourning jewelry, a piece of South Pacific painted bark cloth, a Confederate bowie knife to gold Etruscan Revival earrings, a 1913 magnifying glass commemorating the Romanov dynasty, a 1953 All-Star baseball, a Beatles album cover, a John LaFarge drawing, all retrieved from their attics and garages, old cabinets, trunks, and forgotten drawers. "My grand-mother kept this in her dining room cabinet for years," we hear, or "My great uncle left this in the attic when he got married and moved, and we've had it ever since."

And then we are told that such an object is scarcely worth the weight of the Styrofoam it has been packed in, or that it is worth tens of thousands of dollars, and in both cases the discussion between owners and appraisers considers both monetary and sentimental value. Certainly the drama of the exchange lies in part in finding that an ordinary ceramic bowl or turned table of little known value will now secure a comfortable retirement. But I would guess that an equal amount of drama lies in the narrative, the sure sense of theater created by the breathless anticipation of the owners, the academic dryness of the appraisers as the conversation progresses from where the item was made and how and by whom, to who then found it or bought it and saved it and loved it. Or didn't. And will it be kept or sold? It is difficult not to be captivated by these tales of ordinary things.

It is hard to watch such a program and not come away with a belief in the opera of the inanimate – the quiet fatigue of an old quilt, the spiritual conviction in a perfectly woven Navajo rug, the solid persistence of an eighteenth-century wooden spoon. I once saw an exhibition of paintings by John Singer Sargent that came to this point in a different way. Sargent was the most celebrated portrait painter of his time, yet as the curators suggested, in many of his works the action resides almost completely in the objects; a woman is sleeping on the grass in the garden, but the larger part of the canvas is taken up by her skirt's voluminous folds, creased almost violently. Or the languor of an afternoon tea party is undercut by the cloth that lies awry on the table. The artist seemed to be familiar with the idea that sometimes objects tell stories more eloquently than people.

Certainly those stories have a new resonance at a time when the ephemeral accosts us from all sides; physical objects engage us in a different way today. With our cell phones, e-mail, and assorted forms of wireless communication, the elusive corridors of cyberspace have whetted our appetite for what we can touch, hold, taste, see. In the virtual age, the sorcery of the physical has intensified. We become attached to objects out of sentiment, perhaps, or for their symbolic value – a wedding ring, a grandmother's quilt, an old fountain pen, all of which may commemorate personal history. We seem to accept the idea that things have a life of their own. And that acceptance is the beginning of having an emotional relationship with inanimate objects. Whether the object of one's affection is Marilyn Monroe's chipped kitchen table or a birch plywood table by Alvar Aalto, we seem to have developed a psychic intimacy with our stuff.

But objects make their emotional appeal to us on a much broader cultural level as well. I am certain that their narratives have some tangential connection to our contemporary design literacy. We live in a time when what is called "design" has achieved a certain market status. In *The Substance of Style*, Virginia Postrel argues that the ubiquity of town design review boards, Starbucks coffee, Alessi toilet brushes, and the styling of everything from sneakers to kitchen appliances to computers suggest that design is newly available not just to the elite, but to all of us. Certainly there is no arguing an increased attentiveness to design, but the actual meaning of design seems to vary.

One general and reductive way to define design has simply been to say it is about communication. Another conventional notion of design is that it is where technology, art, and culture converge to solve the problems of everyday life. In the late 1990s and early 2000s, Chrysler sponsored annual design awards, selecting objects for their recognition of color, technology, and craft. Design, the copy for the competition read, is "where form and function meet. The perfect marriage of passion and precision. Memorable designs are those whose very appearance exposes their excellence. And whose performance and engineering only intensify it." Postrel goes a little further in suggesting that emotional content is also essential to how something is designed: "Design provides pleasure and meaning as well as function, and the increasing demand for aesthetic expertise reflects a desire not for more function but for more pleasure – for the knowledge and skill to delight our senses."

Even the *Harvard Business Review* recently went so far as to suggest that businesses are realizing that "emotionally compelling" objects and services are what will distinguish their offerings today.

As a citizen in an age of design literacy, as we all seem to be, I am interested in these assorted definitions, these phrases and verbal wranglings intended to capture the meaning of something obscure. It seems long overdue that the emotional content of objects, of buildings, of places be considered part of the equation. Because at the same time that I have been writing about design – about why mattresses are getting larger although we all seem to sleep less, why vintage rotary phones are finding a new market, or why skateboards can be an accessory to urban experience – I have also watched my two sons grow from infancy through childhood and adolescence through their teenage years. Witnessing the way in which they take ownership of their things has shaped the way I think about design. I have been reading about smart refrigerators for years, and about how what today are called "infopliances" will reinvent the way we not only prepare meals but also communicate with one another. But when my son Luc used the refrigerator as a place to store his dog's fur, it inevitably recast the way I thought about the function of this appliance. When his brother, Noel, came home from school one afternoon with the news that he had been taught how to apologize in a class called Home Careers, it shifted my thinking about how kids learn about domestic tasks. And this is just the kitchen.

Because such moments in the domestic life of my own family encouraged me to investigate how ordinary things are shaped and used, it gave me all the more pleasure to realize that the same quotidian family life, at times, causes things to be the way they are: a vegetable peeler takes new form because a man is trying to help his wife, who has arthritis; a lawn chair comes into being because another man wanted a place for his family to sit outdoors;

and a father improvises a little snowboard so his daughter can slide down a snowy hill. A divorced, middle-aged woman suggests that housewives will buy brightly colored plastic food containers if they are sold at parties, and Tupperware enters the American kitchen. More and more, I found that the ideas, uses, and functions that came up time and again in our domestic life, whether they involved a backpack, camera, or phone, were rarely the same ideas I encountered in design theory or literature, much less in the marketing material purporting to make these objects appealing, necessary, desirable. How they are actually used and the details of how they inhabit our lives often follows a separate narrative.

More and more, that material narrative seemed worth following. Do the concerns of an engineer at GE who is designing a smart fridge ever intersect with those of a child who is considering how to store his pet's fur? Medicine cabinets are a common domestic object that has been growing in recent years, and the shape and size they take reflect values about home security, healthcare, and how we take care of ourselves. At the same time, the medicine cabinet is a potent symbol for contemporary artists like Damien Hirst, who constructed an entire pharmacy of imagined remedies. Is this pure coincidence, or do his reflections on the cabinet share anything with those of houseware designers who are producing their super-sized versions? And in this age of streamlined, accelerated communications, why did my neighbor John Scofield build a FedEx delivery box in the form of a Roman temple? If one is interested in design and in how things take their shape in the physical world, these narratives seem worth following.

In the end, it would be preposterous, of course, to draw grand and universal conclusions from the experience of two boys growing up in the Hudson Valley. Yet at the same time, it is difficult not to admit that their things tell a story not only about them, but also about the rooms, the landscape they inhabit. The strollers and cereal boxes, phones and cameras, and all manner of other ordinary domestic objects that have been their accomplices in everyday life compose a profile of their lives and time. The backpacks my kids use today have a universal appeal to middle-class American teenagers; they are accessories to their way of life, a form of psychic prosthetic for an entire generation. It would never occur to them to use tote bags made in the shape of AK-47 automatic rifles, though such bags are sometimes used in South Africa, where the murder rate is ten times that of the U.S. Such are the revelations of simple things. Our objects signal who we are.

But in recent history, it's not the catalogue of specialized backpacks, nor the volume of goods sold on eBay, nor the revelations of *Antiques Roadshow*, nor pictorial essays on ethnic curiosities that have said the most about our emotional engagement with the inanimate. Rather, it is something that occurred in winter 2004. A minor scandal broke in New York City when it was discovered that several FBI agents had taken an assortment of physical objects from the Staten Island landfill that held the debris from the World Trade Center. Some of the agents had spent long, painful months sifting through the rubble in an effort to collect further evidence of the particulars of the towers' collapse, and as mementos they took such varied objects as a globe paperweight, pieces of metal and concrete, an American flag. Outrage from some of the families followed, the sentiment being that the agents at best were removing evidence from a crime scene, and at worst, robbing graves.

The anger seems misplaced. It is only human to look upon objects and artifacts, though they may be inanimate, as witnesses to human experience. And it is not difficult to understand the agents' desire to have a small, physical piece of the debris. Only a few miles away at the New York Historical Society, an exhibition titled *Recovery* seemed to operate with a motive similar to that of the FBI agents. On display were objects found in the 1.8 million tons of debris at the Fresh Kills site – jewelry, coins, photographs, bits of demolished fire trucks, pieces from the planes, keys, a plaque for an elevator door, a piece of an airline seatbelt, all of them souvenirs of catastrophe. The items were selected with the utmost deliberation. Would it be sensationalizing to include bits of aluminum from a hijacked plane? In the end, Mark Schaming, the director of exhibitions and programs at the New York State Museum, wrote that "the staff at Fresh Kills began to see museums as a way to preserve not only the objects, but this story as well. Museums were giving something back – their work as documented history....These objects will speak to generations about the World Trade Center."

Both the curators of the exhibition and the agents have been called souvenir hunters, but I would imagine Pablo Neruda's observation that many things conspire to tell the whole story applies equally to their enterprises. The desire to preserve tragedy through objects seems less exploitative than an effort to assimilate unfathomable disaster through the familiar. Why else would Amadou Diallo's bullet-riddled doorway be auctioned on eBay? Why else would a man in Mississippi amass a large collection of artifacts related to slavery, an inventory of our inhumanity that includes shackles, whips, branding irons, and sales deeds? While such enterprises may be charged with sensationalizing tragedy, it seems more likely that they reflect the common impulse to observe human history through a culture's objects. The resonant narrative of physical objects elicits a universal response. To be a souvenir hunter is to be human.

The red glazed terra-cotta vases of fifth-century-B.C. Athens are prized not only for their graceful shape and the technical skill that went into their painting, but also for their figurative decoration, which tells us volumes about the way those early Greeks lived – everything from how they ate and conducted themselves in the rituals of ordinary life to how they fought their wars and worshipped their gods. As there are few written historical documents from that time, the red figure pots are the vital evidence. Today our books, photographs, films, magazines, and all types of digital technology record our time. But it seems clear that despite all of these, the resonance of simple domestic objects persists. They, too, are witnesses to their time. Whether it is a smashed messenger's bicycle from Fresh Kills or a horse with a spear in its side inscribed on a red clay pot twenty-five hundred years ago, both convey the fragments of a story. The impulse to measure human experience through the things we can touch, see, and hold comes naturally. We are all part of the market for collectibles.

Like my sons maneuvering the wooden box in their hands, we demand participation from *things*. We ask them to be our witnesses and accomplices. I am certain this is at the heart of design. How things are designed may well be about "performance" or "engineering" or "excellence" or "communication." But I think most of all that the way things take their shape, form, size, color gives us a sense of measure in the physical world; it assures us that the world accommodates us and that we, in turn, can accommodate it and what it brings. That mutual reassurance can be at work in the way a backpack folds into one's shoulders and back, the way a comfortable desk offers a view of the world, or the way the handle of a vegetable peeler conforms to the human grip.

If there is a message, it is that things in this world can find a fit – the way the coin fits in the drawer, the drawer fits in the little wooden box, and the box itself fits into a child's hand, allowing him to believe in the transformative power of the physical world because we can make it a place where mystery, logic, and pleasure coincide. And I wonder if this is the hidden coin, the ability and inclination we have to persuade inanimate objects to be our partners in experience.

many things conspired
to tell me the whole story.
Not only did they touch me,
or my hand touched them:
they were
so close
that they were a part
of my being,
they were so alive with me
that they lived half my life
and will die half my death.

from "ode to things," PABLO NERUDA

THE OBJECTS

the STROLLER

We generally assume our objects belong to us, and generally we are right. But there are times in life when we belong to our objects. In moments of extreme need, or when desire leads to helplessness, the precise details of ownership may sometimes blur, and we can then find ourselves in the possession of our things. Extreme need, desire, and helplessness are all circumstances common to infants, so it only makes sense that they belong to their strollers more than their strollers belong to them; at an early age we are introduced to the conditions of reversed ownership, and insofar as this object is one of the first that mediates and governs one's entry into the larger world, it's a potent experience.

Besides, humans being the contrary creatures they are, the things we desire often accommodate and generate their own paradoxes. Stiletto heels are shoes that hobble the foot; sleeper sofas do neither very well; and car alarms have spawned their own epidemic of intrusion very different from the one they attempt to prohibit. Contradictions of the object world abound, but it is my suspicion that they appear with most regularity and consistency in things designed for children. That is probably as it should be because there are few tasks on earth laced with such ambiguity as raising a child. From the moment an infant is born, everything a parent does is at once geared toward sheltering and protecting and keeping the child from harm's way; and at the same time, it must all be equally geared toward making the child independent. Love him so he'll forever feel safe and nurtured, but above all teach him how to leave. Children's products, by nature, must embrace and accommodate these inherent contradictions, confront this complexity with the appearance of innocence, good will, nurturing. Certainly that is the case with the stroller; with its twin aspirations of safety and mobility, it introduces the infant to the notion of paradox.

Like most other things, children's furniture and equipment observes its own cycles of fashion. Playpens, once considered necessary furnishing for toddlers, allowed kids to play safely while mom was otherwise engaged. Long out of favor, they are seen today as little prisons called into service only by a parent whose need for control is matched by disregard for the process of a child's cognitive growth, a contemptible piece of equipment that limits a child's mobility, prevents the exercise of curiosity, and answers instead to parental convenience. Today, of course, it is mobility we are after. In spite of a flat birth rate and a lean economy, the sale of strollers has been steadily on the rise in the U.S., and their design reflects growing consumer interest and expectation.

Safety, durability, maneuverability, weight, and ease of use have all figured in the redesign of conventional strollers, which increasingly seem to be some kind of hybrid between SUVs and exercise equipment. Among the most popular models in recent years is the Bugaboo Frog stroller designed collaboratively by Max Barenbrug, an industrial designer in the Netherlands, and Eduard Zanen, a doctor. Lightweight and with a seat that adjusts to three positions, it has two small swiveling wheels for city maneuvering, and two large terrain wheels for off road. It can convert to a two-wheel position to be pulled on the beach or snow, has a reversible handlebar and a seat/bassinet, enabling the child to face either direction. And its styling reflects its contemporary functionality – with black rubber and fabric seat that comes in red, gray, or aubergine, it is solid, streamlined, contemporary. There is nothing frilly or frivolous about this baby accessory.

No surprise, then, that it has become an early childhood status object with a requisite appearance on *Sex and the City*. As one reviewer wrote in 2003 in the *New York Times*, "Designer diaper bags do nothing for me. Sippy cups are for kids. But the $700 Dutch engineered Bugaboo Frog had me rolling down the path of conspicuous conception. Maybe it was the twelve-inch all-terrain tires or the squishy grip bar or the fact that the Bugaboo steered

more like a Porsche than a pram, but here I was, wheeling an empty stroller through a grocery store for the adrenaline rush. I wasn't alone. From Central Park West to Santa Monica's trendy Montana Avenue, the Bugaboo is the chariot of choice."

Even the names of the latest generation of strollers suggest these may be the baby's first vehicle for the rough terrain of childhood. Once the names of traditional English perambulators reflected a preoccupation with royalty, as though the Queen, the Princess, the Duchess all might just wheel the infant into an early association with the monarchy. Royal residences were tapped into as well – possibly a pram called Windsor or Sandringham might confer similar real estate upon the child in adulthood.

The contemporary stroller observes a different hierarchy of taste: the Evenflo Journey Premier, the Expedition Double, the Jeep Liberty Limited all take their nomenclature from the automobile industry more than the British monarchy, and like the sports utility vehicles that appear to have spawned them, the strollers evoke qualities of fitness, mobility, strength, endurance, independence – though to whom these qualities belong is not quite certain as they are unlikely to have developed in a three-month old. With dads taking a greater role in childcare, small wonder that oversized SUVs have become the new model for strollers – though it goes without saying that like most SUVs, many of these strollers are likely to be used more in the suburban avenues of malls and food courts than in any more demanding rural terrain.

And while they still may be called strollers, strolling is about the least of what they do. A nostalgic and outdated term that once suggested when we took the baby out for a walk, it evoked a gentle and easy promenade. The strollers we use today may be called that, but equipped with shock absorbers, knobby tires, titanium frame, and nuanced suspension and steering systems, they seem more suited to a sprint along a rocky mountain trail.

Then there is the hybrid stroller, those models that morph into car seats and airline booster seats, and one designed by a mother who is also a flight attendant that can attach to a rolling suitcase so the child can be towed along with the luggage.

Yet for all their presumed commitment to mobility and fitness, there is a perception among some child psychologists that the prolonged use of a stroller may, in fact, hinder the development of those very qualities. Many of these vehicles are marketed for children from infancy to four years, and there is concern that increased use and dependence on one may limit a child's independence; that once a child is able to walk, being confined to a stroller diminishes curiosity and restricts the sense of exploration necessary for development. At three and four, children want to investigate the world, and the tendency to keep them in a stroller may stifle that instinct. Let the child walk, run, explore the world beyond the confines of the stroller, such advice begs.

But parents, especially those in cities, beg to differ. The real service of strollers, they say, is that while offering parental convenience (many must also double as shopping baskets and storage bins), they provide a small, safe place for children to sit in a chaotic and unpredictable world. And that sense of safety may be extended when viewed through the child's eyes. Jennifer Carpenter, a designer and young mother in New York, sees the stroller as more "than a convenient way to move a toddler and carry all your bags; it can be a total baby haven." When her son is in the playroom in her apartment building and feels intimidated by an older, larger child, he resorts to the stroller; for him, it is not simply a vehicle of mobility, but a place that confers its own comfort and security.

That comfort and security are, of course, elusive, changing from generation to generation, culture to culture. Cataloging cultural preferences in strollers in *Metropolis* magazine in 1993, Michelle Herman observed that while Americans favor strollers that can be adjusted so the child can face the parent, this dual positioning is of little value elsewhere. The large wheels of German strollers don't swivel. The Japanese prefer lightweight strollers, while those sold in England are rarely equipped with sunshades. Strollers used in France don't have rain shields, which are thought to be overprotective. In the late nineties, a mother from Denmark living in New York City was arrested for leaving her small child in a stroller on a sidewalk while she was indoors in a restaurant. Such practices are routine in Denmark, where it is believed that exposure to cold winter air is bracing for even the smallest children and where there is a low incidence of street crime. In New York, however, her action was perceived as criminal negligence.

Even the object itself can convey associations that seem antithetical to its intent. In summer 2004, strollers joined guns, explosives, bottled water, and umbrellas as items banned for reasons of security when the Republican Party held its convention in Madison Square Garden. That a stroller, an object designed and devoted fully to the safety of the innocent, could be construed as a threat to public safety says something about how our readings of things can be reversed entirely depending on the context of their use.

All parents want to hold their children close; all parents want their children to be safe. But what actually provides safety is more elusive. And how we give our children knowledge and experience is not something that can be easily defined or quantified; if it could be pinned down for even a moment, it would change. The playpens favored by a previous generation seemed to indicate a fondness for children learning to engage themselves quietly in a room; the contemporary generation of strollers puts a higher value on getting out into the world and on the conviction that early childhood is an all-terrain experience. In the eyes of one parent, leaving a child in a stroller on a snow-covered street confers resilience and a capacity for solitude, while to another it is simply thoughtless neglect. I am certain there are absolutes in child rearing, but I am equally certain they won't be found here; even in adulthood, there are few final and satisfying answers on whether sitting in a room quietly or being outdoors fully enlisted in physical activity is the more genuine engagement of human experience.

When my own sons were very small, I would take them for walks on our country road in a small gray twin stroller that had been picked up at a yard sale. This was not a Journey Premier, nor did it bear any resemblance to the Expedition Double. There was nothing very substantial about it; with a gray fabric seat and back, a metal frame, and fabric canopy, it was a delicate armature for their early outings. Like any new parent, I found our quiet country road a catalogue of worldly hazards. Even with the canopy adjusted, the bright summer sunlight seemed to stream onto the boys' small, white faces. On one side of the road was a marsh occupied by snapping turtles. That summer, their offspring could often be found near the road, and although they were no larger than bottlecaps, with their gray armored shells and pointed, prehistoric tails, they conveyed a sense of menace, albeit in miniature.

Imagined dangers were ever-present as well – a teenage boy behind the wheel of a car, possibly, or a truck suddenly swerving to avoid a deer. And our neighbor had a German shepherd who was predisposed to fury and would charge the stroller at our approach, barking ferociously. I learned the surest way to avoid confrontation was just to continue walking without trying to appease the dog. My children never so much as blinked. One morning, my neighbor ran outdoors in her pink bathrobe when she saw us coming and handed me a brown paper bag of marshmallows. "Here," she said, "he likes these. Give him one or two of them, and he'll calm down." I did, and he did.

I think about all the walks we took that spring and summer, about the tiny snapping turtles and the loud angry dog, and it occurs to me that the stroller, while being all about safety, also offered a guided tour to an index of earthly dangers, real and fictive. That summer was a time when even the smallest snapping turtle conveyed a sense of danger, and I try to envision taking those walks with my small sons in a Bugaboo Frog from the Netherlands with its wide, sturdy tires, or the Balmoral, a high-end luxury pram from Silver Cross that is an extravagant construction of lacquered steel, leather, and chrome. Maybe a three-wheel sports utility stroller would have conferred a speed and maneuverability that would have seemed useful at the time. Would any of these vehicles have made a difference, I wonder. Would they have provided the framework for safety that I wanted for my sons? Would the country road have seemed safer to me? Would any of these strollers have given much resistance to a speeding car or an angry dog? And I recall that safety, that particular summer, if it came from anything, came from a brown paper bag full of marshmallows.

Safety in children's equipment is nothing to be glib about. There is every reason to rate strollers for their safety features, their five-point harnesses, reliable parking brakes, the ease with which they fold up, adjustable-height handles, and all the other features this generation of strollers has been designed to include. But safety itself is more difficult to outline and catalogue, and the dangers of this world even more difficult to define, much less anticipate. While it is hard to fault a parent for hoping that a three-wheel stroller with heavy-duty suspension is the device that will keep his or her children secure, or a $2000 titanium stroller is the equipment that will hold them safe from harm, protection itself is surely a more complex question, its most reliable accessories more difficult to determine.

the CEREAL BOX

Reading and eating are an infallible combination. Everyone knows this, but kids know it especially well, because when we are children, words and food are permitted to congregate on the table effortlessly. What is alphabet soup if not the assurance that reading and eating can be one and the same? It is only as we become older that we are burdened with the insidious notion that eating is a social, rather than literary, activity; that words and food become separate pleasures; and that typography on the table is perceived to be a literary digression that intrudes upon the sensual pleasures of eating. If we find it at all, it is limited to the occasional exquisite wine label or infrequent place card.

Which is a shame. Who is to say that the palate and the imagination cannot be provided for at the same time? Not whole books necessarily – bringing books to the table can be cumbersome and impractical. Still, I am certain that there is a place at the table for type, notes, fragments of text, syllables of thought that go beyond the occasional fortune cookie. Maybe it is because they are both agents of sustenance, but food and words enjoy a natural alliance; as in the best collaborations, each elicits the strengths and pleasures of the other. One of the more memorable meals I have had recently was a picnic on a July afternoon. Midway through the luncheon, a tiny note unfurled itself from a salad of chicken and walnuts. The words, indecipherable, could have been the fragments of the recipe, or possibly even the remains of a grocery list, but the fact was it didn't matter. The possibilities of the message transformed the meal, not just into an exercise of absurdity, but into a consideration of the infinite promises of the unexpected and unknown.

As strange as it may have seemed to find a note hidden in the chicken salad, it reminded me of an even more astonishing moment that occurred some forty years earlier at my grandparents' kitchen table in Massachusetts. As I opened a cereal box and tipped it toward my bowl, an emerald green plastic ballpoint pen shaped to resemble a feather quill pen fell out of it. The notion that unanticipated gifts might drop out of a cereal box was a thrill to the imagination, and I learned then what most kids sooner or later find out: the breakfast table can serve as a library, a place to read, and, on this particular morning, a place to write; with that came the knowledge that sometimes a place can be defined not by space or furnishings, but by what you do there.

It was not until decades later and after my sons were born that I learned this alliance between food and type had a name. The manufacturers of breakfast cereal have dubbed it the "eat and learn" phenomenon. I am certain that one reason breakfast remains a favorite meal for so many kids is because not only are they allowed to read there; they *must* read there. Reading is permissible at breakfast. In the early morning, silence is accepted over sociability, and even adults read the morning news over a cup of coffee at breakfast. Quiet is not only accepted, but at many breakfast tables it is outright cherished.

Our own breakfast table, lavishly outfitted with cereal boxes, was surely my sons' first literary experience, and if the boxes didn't contain unexpected writing utensils, they were still fantasy furniture for the child's brain, constructed to challenge them on any single morning with their profusion of fact and fiction, the biographies of athletes, crossword puzzles and word games, the brief cultural history of ghosts, the scientific construction of food pyramids. With all their offers, information, contests, and prizes, it is small wonder that it's where kids grasp the fact that written symbols can be fastened together to represent the names of people, things, and even abstract ideas.

Chris Capuozzo is a graphic designer in the New York City firm Funny Garbage, and he and his partner, Peter Girardi, cheerfully admit that cereal boxes may have had something to do with their adult careers in graphic design; even the names – Count

Chocula, Boo Berry, Fruity Pebbles, King Vitamin, Frankenberry —
make for a phantasmagoric mythology where American commerce
and childhood fantasy converge. "It was both the images and the
words," Capuozzo says. "The logos were bright and fun and
healthy; it felt like the experience of munching out itself. Letters
with bright colors had an exaggerated presence...this really was
the taste of cartoons we would be watching, like if you could pull
off a piece of Fred Flintstone's house to eat, it might taste like
Fruity Pebbles....Eating this cartoon food had a certain kind of
psychedelic aspect to it." Girardi adds, "The combination of characters and decorative typography was a huge influence; later on,
when I became a graffiti writer in New York, it was a kind of natural extension of what I loved about cereal boxes."

With that kind of resonance, it's not much of a stretch to
suggest that book jackets and cereal boxes reside in parallel universes.
(These universes may even occasionally intersect. In spring 2004,
*Free Prize Inside: The Next Big Marketing Idea* was published with
a cover resembling a cereal box, with the image of a superhero and
bowl of cereal. The book also had a limited release in which it was
enclosed in a real cereal box.) And while the purchase of a Count
Chocula box from the seventies may not have quite the literary
cachet of a first-edition Faulkner, cereal boxes have still generated
their own subindustry of collecting: the 1993 Michael Jordan silver
commemorative Wheaties box was selling ten years later for a mere
$19.99, but a 1969 promotion for Nabisco's Wheat Honeys that
featured characters from *Yellow Submarine* was being sold as a
Beatles collectible on the Internet for $2000.

If you accept the notion that the breakfast table is a
reading room for a lot of kids, it's no surprise that some cereal and
snack manufacturers, enlightened to the idea (not to mention the
infinite marketing possibilities) of reading at the table, are entering
into licensing agreements with publishers to put out "snack books."

My plastic green quill pen conveyed a sort of transcendence: Write anything! it suggested, and its very presence seemed to embody a kind of alchemy. Its magic, other than its origins in a box of Corn Flakes, was derived from its implicit suggestion that the words could be imagined by a child at breakfast. The pen was at once old and new, its antique form rendered in bright plastic; as such, it was probably the first object I ever owned that taught me about how a single thing can straddle centuries and sensibilities. Free Prize Inside indeed.

As practiced today, however, the "eat and learn" phenomenon leaves less to the imagination. *The Oreo Cookie Counting Book*, for example, teaches toddlers more conventionally and more methodically to count down to "one little Oreo...too tasty to resist." Other such literary mergers between HarperCollins and Pepperidge Farm Goldfish and Simon & Schuster and Sun-Maid raisins offer an assortment of mealtime activities thought to advance hand/eye coordination, counting skills, and vocabulary. In its commitment to literacy, Cheerios has gone so far as to package small books with the cereal. The breakfast cereal launched its Spoonfuls of Stories series during National Children's Book Week, November 18–24, 2002, with such titles as *Salt in His Shoes: Michael Jordan in Pursuit of a Dream*, a testimonial by the basketball player's mother and sister, and childhood favorite *Alexander and the Terrible, Horrible, No Good, Very Bad Day* by Judith Viorst. Specially sized to fit in cereal boxes, the books are said to be directed toward low-income families who have no books at home for kids.

There is no denying that the idea of snack books has limitless possibilities. We know that eating and reading are a tried-and-true combination; it is hardly a snack book, but *Remembrance of Things Past*, arguably the greatest work of twentieth-century fiction, was inspired by nothing more than a cup of tea and a small cookie. And surely the adult market is more than ready for the eat-and-learn phenomenon. Could gourmet cuisine not be used to further adult literacy by pairing certain foods with specific writers – sliced hams, cod roe, toasted cheese with Samuel Pepys or a croissant with Colette? Later in the day, one might even engage in a drink-and-learn phenomenon – rum cocktails with Hemingway, gin and tonic with Cheever.

Despite the commercial possibilities, the eat-and-learn phenomenon remains confined to the children's table. Yet for all its promise, questions persist. While it may be hard to argue with such advocacy programs for literacy, the question remains whether these foods, known in the industry as "branded consumables," promote eating and learning so much as they do a more insidious conjunction of kids, a high-sugar, high-fat diet, and rapacious marketing campaigns. (Surely that's one of the innate messages of *Free Prize Inside*: it only makes sense that if a book is going to be designed to resemble a cereal box, it is going to be a book about marketing.) A 2004 report by the Henry J. Kaiser Foundation traces a direct connection between the growing rates of childhood obesity and advertising directed at kids; and while the research does not target snack books, it calls into question the need for and legitimacy of media assault – everything from TV ads to the graphics of food packaging – on children's eating habits.

Besides, such books may be redundant. Who needs them? What appears on the kitchen table already has a complex narrative all its own. As Chris Capuozzo says, cereal boxes are by nature "vibrant and intense. They speak strongly; it's really a tone that the graphics have, this hyperreal life force. The characters and graphics have their own kind of superpowers. It's part of knowing what kind of presence a graphic can have to be really powerful." Even General Mills might have underestimated the reach of the cereal box when it tucked free CD-ROMs of the Bible inside some of them. Certainly the subsequent narrative was more subtle and nuanced than anything a Froot Loops interactive journal or Cheerios playbook might offer.

In the summer of 2002, General Mills put together a promotional CD-ROM with computer games – Clue, Lego Creator, Carmen Sandiego Word Detective, and a bonus free Bible – prompting an editor at *Publishers Weekly* to remark, "Wow. Breakfast cereal, computer games, and the Bible thrown in too. What a cultural fruit salad this makes." While the Bible itself was not mentioned in the promotion, it could be accessed on a selection marked Home Reference. A unique coalition of objectors formed almost immediately, testing the assumption that if 90 percent of the country is Christian, surely a free Bible would be welcome at the breakfast table. Disney Interactive, which had licensed software for a game based on *Who Wants To Be a Millionaire*, had determined the inclusion of the Bible too controversial and withdrew its name from the marketing package. Meanwhile, other religious organizations voiced their own discomfort, and General Mills found itself making a public apology for the promotion almost before it started – though not until twelve million boxes had reached grocery store shelves.

The cultural fruit salad gets a new mix with the Wheaties box, certainly one of our more resonant domestic icons. With its illustrations of athletes something akin to *Time* magazine's person of the year for kids, the bright orange cardboard box is one of the more significant monuments on the landscape of the kitchen table. The cereal celebrated its seventy-fifth anniversary in 1999, and as part of its celebration put Muhammad Ali on the box front. Ali was at the height of his career in the mid-sixties; he won the gold medal for boxing in Rome in 1960 and went on to become the first man to capture the world heavyweight championship three times. That was also when he joined the Nation of Islam and changed his name from Cassius Clay.

When asked why it took General Mills thirty years to recognize Ali's place at the breakfast of champions, a marketing manager for Wheaties told the *New York Times*, "I think it was a culture thing. At the time when the greatest boxers were boxing, the people weren't ready to accept them on the cover of the Wheaties box." He added, "For the character and equity of what he represents, [Ali] was an easy decision for Wheaties now." "Turn your morning into miles," Kellogg's All Bran suggested in its ad for a mileage program with American Airlines, but the facts suggest we haven't come all that far. The most reassuring and visible personas in the breakfast cereal aisle continue to be the Quaker on the oatmeal packaging and Nabisco's cheery, coffee-colored cook offering up steaming bowls of cream of wheat. General Mills went so far as to put Smash Mouth and Boys II Men on its packaging for French Toast Crunch, but they were on the back of the box with an offer of free CDs.

Yet the most haunting tabletop graphics may not be in the words or images found on the cereal box, but on its necessary accompaniment, the milk carton. That relic from the mid-eighties, the cardboard milk carton printed with the call for abducted children, came imprinted with all the essential information – the blurry photograph, date of birth, sex, hair color, eye color, date missing. And attending the grainy, imprecise image, simply the word "Missing." Who could have thought that such a flimsy piece of cardboard might express our deepest fears, or that a milk carton, with all is implications of sufficiency and good health, could bear the imprint of unbearable absence?

Child psychologists have long since determined that these images of missing kids were unnecessarily disturbing to small children, and those cardboard cartons were replaced with translucent plastic jugs, their graphics reduced to a small label cataloging fat content. Recognizing fat content may be important to maintaining good health, but the evolution of those graphics seems another indication that self-interest, in this case a cultural mania for measuring cholesterol, has replaced a broader recognition of human community. It must be another one of those culture things.

The artist Jeff Koons, clearly no stranger to the eat-and-learn phenomenon, once observed that cereal boxes were more interesting than most contemporary art. "I've always enjoyed Cheerios," he said. "But all the breakfast cereal boxes are exciting. The reason is they're trying to pump energy into people in the morning, make people feel good about the day." His comment is facetious only in part, and I suspect that as much as Koons is enamored of the box's presumed intent to generate such good feelings, he is equally smitten by its cultural fruit salad. Because if the kitchen table is where kids learn about sustenance, sharing meals, and familial civility at large, it is also where they learn about disposability, and food that is fast, easy, and cheap – not to mention design, commercial branding, the politics of packaging, religion, race relations, and the fact that biblical figures, basketball celebrities, Muhammad Ali, and missing kids are all part of the dialogue.

Eat and learn for sure. For generations now, the cereal box has been a participant in the conversation at the table. But where I once had an emerald green plastic quill pen, my sons have branded consumables. These lightweight, disposable boxes convey attitudes and values that are anything but. Because most of all, implicit in the graphics of the kitchen table is something more basic – whom we invite to sit with us at the table. And when. And why.

The BACKPACK

I once read an anthropologist's insight that women love to carry things. No matter whether it is a farm woman in Kenya balancing a woven basket or water jug on her head, a Japanese woman with her *furoshiki*, or contemporary urban Western women with their purses, clutches, and shopping bags, women, he observed, are more comfortable in their lives when they are carrying things. It is to my eternal regret that I cannot remember who said this, but I have probably forgotten his name simply because the notion seemed so misguided. If he had just substituted the word "child" for "woman," there might be some truth to it.

Because children *do* like to carry things around. First indulging in the impulse to collect and carry is as important as the other more obvious milestones children experience as they become independent beings. Along with the first haircuts, words, steps, and sentences is the emerging awareness that the things one collects and carries help compose identity. In our family, this knowledge arrived in the nylon backpacks my sons carried when they were five. Bright yellow with red and green straps, their owners' initials printed in blue, the packs also contained scraps of paper, bits of crayon, leaves, twigs, superhero action figures, coins – a random assortment of what small boys might consider, on any given day, essential accessories.

That my sons, along with countless other children, find such allure in backpacks only stands to reason. As infants, many children were carried around in some variation of the backpack – front packs and slings, back packs, multiuse packs that could be converted to little cribs and sleepers – and I suspect there must be some lingering subliminal memory of this experience. That a backpack is the place where you carry around what matters in life has been imprinted upon them, and carrying around their own small packs confers importance, responsibility, and that quality small children love and desire most, authority.

With all their pockets and zippered compartments, backpacks also embody the idea of legitimized secrecy. A friend of mine recalls that her son at four would go nowhere without his backpack, inside of which was a small plastic trout. Such an arrangement answered to an early taste of privacy, not for oneself but for one's things, for the objects of one's affections. And in whatever elusive way we identify with the items we own, is it far-fetched to imagine that ideas about adult privacy may have their roots in the way a young boy carefully arranges the placement of a treasured plastic trout in a backpack?

That the things you carry on your back are *necessary* things follows a physical and emotional logic. A backpack connotes safety, self-reliance, rugged independence – all qualities we like to think of as uniquely American. Less frivolous than a shoulder bag or purse, it implies that what is inside is needed. A good backpack provides a sense of assurance; it gives credibility and a certain gravity to this impulse to collect and carry. With its legacy in hiking, camping, and climbing – indeed, survival in the wild with only what you can strap on your body – there is something about a backpack that suggests one is efficiently outfitted, prepared for whatever is to come. One is equipped with the essential.

Not that what we perceive as essential isn't in a constant state of flux. A generation or two ago, a backpack outfitted with the essential might have been a canvas rucksack holding a Swiss Army knife, rain gear, a nylon tent, and other accessories for coping with the vagaries of the natural world. But the prevalence of the backpack today suggests that our list of essentials has expanded and might include, for example, the D-Air jacket designed by Dainese Technology Center in Molvena, Italy. The electronic airbag, designed for motorcycling and other extreme sports, applies the technology of the automobile airbag to the back. The sensors of its electronic control system determine when it should be instantaneously inflated to protect its user from back, neck, or head injuries. Cultural critics have noted that the pack has a secondary use: in overcrowded places such as the Tokyo subway where there is fear of airborne germs, the pack can be inflated to create space between crowded passengers.

Or it might be a more conceptual exercise, such as the virtual Louis Vuitton pack by the French designer Ora-Ito. Part of a piracy campaign that takes the brand name and logo of celebrated consumer goods and subverts them through the creation of a virtual product through advertisements, the bag resonates with irony. Or it might be Swedish designer Jonas Blanking's backpack, a sculptural apparatus inspired by the tough shell of a beetle's wing. What we need should be simple but often is not, and backpacks take this into account; Blanking's Bohlbee pack, with a rigid exterior formed out of injection-molded ABS plastic and a textile harness, was designed so that bicyclists could carry their laptop computers and cell phones with them. Ergonomidesign, a firm in Stockholm, has designed the Spiromatic, an air canister worn by firefighters on their backs; surely there is nothing more essential to carry on one's body than oxygen.

But the specialization of backpacks seems to proliferate most exuberantly in kids' backpacks. Just as a few years ago there seemed to be a specialized glove for every sport, today there appears to be a specific backpack for each sport, all of them a way for children to feel that they are well equipped, efficiently outfitted, that is to say *prepared* for whatever comes their way. Life is composed by categories of the essential. My sons have a backpack for their skateboard gear, another one strictly for snowboarding, another for their cameras. There is even a mini pack that attaches to the backs of their snowboard bindings and is used for wax, Chapstick, and candy, I am told. I have seen packs with side compartments that fold out, designed specifically for CD and MP3 players. All of which may be a way of recognizing that things, along with people, have a place in life. And who is to say what the essential tools for any of us are? A stuffed trout, a tube of Chapstick, a CD player may simply be the beginning of determining what might be essential.

But it is the backpacks used to carry books, the ones utilized most often by kids, that have caused the most problems. When my sons were in elementary school and junior high, the packs they carried could be twenty or twenty-five pounds, accommodating voluminous textbooks whose size surely reflected continually expanding bodies of information in history, science, mathematics. But the weight of their packs didn't just reflect the growing data in the age of information; it also reflected emerging fears. This was a time when many schools were eliminating lockers. They took up too much space in already overcrowded schools, and besides, they were where kids presumably stashed their drugs and their weapons. School administrations seem inordinately fond of simple answers, and the simple answer here, of course, was just get rid of the lockers.

Without lockers, then, the books were lugged to and from school each morning and each evening. Some forty million children carry their books in backpacks at a time when they are growing, their spines extending and changing. Research has shown that more than half the children who use backpacks carried more than the recommended 15 percent of their body weight in them; 20 to 25 percent is much more common, and that can throw kids off balance.

Children tend to lean forward, and as they're growing the spine should be kept straight rather than weighted forward. One of the worst problems occurs when the pack is slung over the shoulder, an action that can both curve the spine and can make muscles develop unevenly. Finally, if the straps are too thin, they can cut off circulation. Walking around with heavy backpacks can also lead to something called "rucksack palsy," a kind of antique parlance that would have a certain charm were it not for the condition it describes, which is "numbness and often, nerve damage, caused by excessive pressure applied to the nerves in the shoulder."

One reason that we have allowed a generation of kids to haul these heavy packs around may have to do with the books inside them. Perhaps it's part of the reaction against the digital age, but we seem to cherish the physicality of books. And if packs connote the essential, what could be more essential to life than books? Possibly you could argue that these packs overloaded with printed matter are simply taking this idea to the extreme. All the same, we live in a time and place when literacy is low. Between forty and forty-four million adults, some 20 percent of adult Americans, have difficulty with the basic reading and writing skills necessary for everyday life. Figures compiled by the National Center for Educational Statistics in 2000 tell us that 37 percent of fourth-grade students tested below basic reading proficiency. I also wonder if there is something in us that values the sheer materiality of books. If our children aren't reading them, at least they're carrying them around; and if we are having trouble teaching our kids to read books, perhaps we are making some headway by causing their backs to ache from carrying them.

What we commonly think of as "design" only haltingly remedied the situation, which seemed strange when I consider that nearly everything my sons outfit themselves with, from their skateboards to technotextile sweaters to their bike helmets, has been *designed*. Space-age materials, witty graphics, provocative logos, inventive styling are applied to everything from their basketball shoes to their baseball caps. Yet there were some fairly explicit design problems here that had not been addressed, not the least of them being how to construct lockers that afforded some degree of privacy *and* security, where in the schools they could be installed, how to fashion backpacks that distributed weight evenly, and how my eighty-pound boys could go to school every morning toting twenty-five-pound backpacks without hurting themselves.

One backpack, the Rakgear, tried to resolve the dilemma with wider straps and an interior skeletal frame system that organized the space inside the pack while distributing its weight evenly over the back. And in fall 2002, Nike somewhat belatedly introduced a backpack aimed at ten- to eighteen-year-olds. Designed with input from doctors familiar with the cause of back and neck injuries, the BioKNX had a panel that conformed to the curve of the spine and distributed load more evenly, along with padding and thick, contoured shoulder straps. Logos, color, and separate compartments for CD players all cater to teenage tastes.

In the end, however, it wasn't these reconfigured backpacks that made the difference; things take their shape in design studios only in small measure. When change in form is necessary, essential even, design becomes a more elusive, collaborative process. In 2002, Gray Davis, then the governor of California, signed a bill banning textbooks exceeding a given weight. New Jersey and Massachusetts considered similar bills.

But legislation alone is rarely enough of an answer. The year 2002 was also when some sort of enlightenment occurred in my sons' middle school. When they returned to school that fall as eighth-graders, their backpacks lightened considerably. Amid the emerging awareness of the problems caused by overburdened backpacks came the call for the use of digital texts; while they were obvious as a lightweight, easily updated research tool, the scant budgets especially in most public schools made this an astoundingly impractical and insensitive proposal. Instead, many of the teachers simply arranged for students to keep their books at home and to use a separate set of classroom books when in school. In some cases, sheetwork could be copied and taken home.

The Backpack

Change in this case wasn't a product of new technology. It came from common sense, or rather a sense of attentiveness, the kind of watchfulness that is often a prelude to change. Surely the designers at Nike who consulted with orthopedists in reconfiguring the backpack were thinking creatively. But the real change in the form and use of these bags was instigated every bit as much by tired kids, angry parents, thoughtful teachers, responsible school boards, and a growing public recognition of a problem. True innovation wasn't simply about redesigning the backpacks, but about redesigning how kids did their homework, which may be the most important lesson of all. The age of technology has situated us in a realm of infinite adjustments, and often these adjustments can be a matter of minute shifts of attitude as much as with any new form or material. While a sixty-pound child is unable to manage the weight of an eighteen-pound pack, his parents and school can respond to the problem, and the circumstances of its use can be reshaped just as surely as the object itself. Such societal adjustments are surely more significant than the adjustments of straps, frame, and fabric that give a backpack its fit, or any other material alterations that occur in the studio. More important, perhaps, they can come from any segment of the population. Material and social change rarely keep an even pace with one another, but their awkward two-step can have its own peculiar grace, and it is one that sometimes results in genuine progress.

Backpacks have become a virtual prosthetic for most American kids, so it stands to reason that their design and use have both received such scrutiny. When I was in elementary school, my friends and I toted our books around in leather book-bags; modeled on the briefcases our fathers carried, they conveyed – or we at least hoped they conveyed – authority and adulthood. The closest thing I had to a backpack was a multicolored embroidered wool bag from the island of Mykonos. While both its form and embroidery had their origins in Greek folk art, this type of bag was produced more for tourists looking for genuine ethnic craft. Its thick red yarn straps allowed it to be worn as a backpack, which I sometimes did, but its weight was never an issue. The bag had been sent to me by my sister, who was living in Athens that year, and I loved the way its red straps connected me to her and to her experiences in some elusive but important way. Still, it was purely decorative. It carried nothing that aspired to the essential.

The backpacks my sons use share nothing in function or design with my red woven bag from Mykonos; they are not ornamental, nor do they have a connection to anything so personal. Made of water-resistant ripstop nylon with adjustable chest and waist, and articulated-foam shoulder straps, they have moisture-wicking mesh laminate linings and are impervious to all kinds of weather or any of the other sorts of indiscriminate battering they will receive from teenage boys. Their multiple pockets and compartments are zippered, strapped, Velcroed, assuring the inviolability of their contents. The packs are held close to the body, some prosthetic device the exact shape and function of which are ever changing. And in all this efficiency is an inevitable sense of urgency, as though these backpacks are apparatus for dire times.

I can't help but wonder whether the apparent desire to have what is important at hand at all times, the insistence on the purely necessary, is somehow particular to our times. We live, after all, in an era when we are routinely advised to have community disaster plans and household disaster kits. Depending on where we live, the apparent threats range from terrorism to earthquakes to blackouts to mudslides, and we are counseled to be prepared with Go Bags packed with radio, flashlight, food, water, medication, money, and all the assorted other items we cannot do without. The Red Cross offers survival kits online that feature everything from food bars to duct tape and breathing masks. Retail stores are

55                                    THE BACKPACK

at the ready as well: the outlet Gracious Home sells a Ready Kit emergency backpack containing food, water, flashlight, wrench, face mask, and biohazard waste bag.

There is nothing, I know, about a skateboard, fish-eye lens, or CD that qualifies it as survival gear. Yet their placement in the backpack gives these articles an association, some imaginary proximity, to more genuine survival gear. It is rare that any of us know, at the right place and moment, just exactly what it is we need. But we like to imagine that we do and take pleasure and meaning in defining our necessities. Maybe that's why this pack has so lodged itself in the imagination of our kids. While what they consider essential may change by the moment – a leaf, a plastic trout, a book, a laptop, a phone – what seems certain is that they will carry it on their backs. Because most of all, the backpack conveys the reassurance that we know what we need and we have it with us.

the DESK

When our sons were five, my husband and I took them to the North Carolina beach for a week. Although they had been to Cape Cod as infants, this was their first real brush with the ocean, and we went on walks, splashed in the surf, built sandcastles, and watched the dolphins from the little balcony of the house we rented. Many months later, when I asked Luc what he remembered most about our week at the sea, he dreamily replied, "I loved going to the ocean. It was where I had my own desk."

And this was true. We had stayed in a little cabin perched on stilts at the edge of the shore. It was the quintessential beach house, minimal design at its best, equipped with a few primitive and rusty appliances, paneled in cheap knotty pine boards, a frail structure that seemed to shake and shudder with every gust of wind. In the room where the boys slept was an unfinished pine table with drawers – indeed, Luc's first desk.

There is an odd but very real logic to the fact that the ocean and the desk were parallel experiences in my son's life. I thought of this years later, in 2001, at an exhibition at The Museum of Modern Art in New York titled *Workspheres*, which set out to reimagine how people work in what at that time was called with great fanfare "The New Knowledge Economy." Because there I found a desk that, while not featuring the ocean, is equipped with the sky. Created by the Japanese designer Naoto Fukasawa, the desk – or desk system, as such things are sometimes known – allows for three different views of the sky to be projected on an overhead screen. Climate and season can be programmed into the atmosphere of work, and light, color, and imagery can all be amended as on a screensaver. As a designer, Fukasawa is preoccupied with work and territoriality, boundaries both natural and manmade, and how the places in which we work can accommodate and reflect those.

Fukasawa's interest in boundaries is shared by architects, designers, consultants, and pretty much anyone else who has thoughts about how people work. It is accepted wisdom today that the conventional boundaries of the workplace have vanished. Wireless technology has made the eight-hour work day a thing of the past, and the fact that people can work anywhere and anytime has suggested that traditional limits of time and space in the workplace have become irrelevant. No surprise, then, that the desk has invited all kinds of reinvention; and even if they are not outfitted with the sky, other notions about changeability seem to percolate in a new generation of desks. If Luc's desk by the ocean is something office planners long to replicate, so was another piece of his experience. Work being too fluid a process to be confined to any single piece of furniture, "hotelling" was a strategy introduced in the nineties that called for office workers to check into temporary desks the way they might check into a room when on vacation.

Maybe it was the disingenuous premise of the hotel, or the implication that a desk was akin to a vacation suite, but for whatever reason, the desk as a temporary guest residence didn't take hold. In his "Ode to the table," Pablo Neruda wrote, "Tables are trustworthy: / titanic quadrupeds, / they sustain / our hopes and our daily life." Designers and office planners met with more success when they recognized that as part of that trustworthiness people want their desks to be their own, and when they acknowledged the genuine value of permanence, privacy, and ownership. The Wake Desk produced by Haworth in 1996 consists of a small curved metal tray perched on a pedestal. Floating above it is a semicircle of metal from which dangle binders, clips, gooseneck clips, saucers, and little shelves, all meant to accommodate everything from coffee cups to Post-It notes to disks, and all of it suggesting something Alexander Calder might have imagined had he been an industrial designer rethinking how to build office systems.

Likewise, Ayse Birsel's Resolve work station designed for Herman Miller in 1999 is a constellation of surfaces that can

constantly be shifted and realigned. Eric Chan's Kiva table, also designed for Herman Miller in 1999, is a fluid collection of tables, screens, storage space all intended to be moved constantly, as though the fluidity in the form of the furniture is a material expression of the ever morphing ideas and processes of the new information worker.

But such changeability, I would suggest, is an innate characteristic of the desk. Much of what we do when we're working has to do with making conceptual adjustments; a desk is a place where one thing leads to another. It makes sense, then, that where we work often has an association with another realm of activity altogether, and there is a logic to why people are so often drawn to the idea of working at tables that were meant to serve another purpose entirely – why college students work on upended doors and why dining tables, dressing tables, kitchen tables are so often put into service as desks. A friend of mine who is a writer works at a desk that is, in fact, a modified changing table that had been originally built for his infant daughter. Perhaps because one's desk is the place where ideas are shaped and reshaped, it does not seem unlikely that mutability is essential to the character of a desk itself, and the image of a desk as a changing table has an innate resonance.

Security comes into play as well in the design of desks. Considering that "knowledge is power" has become the mantra of the information age, it is no surprise that something called the PowerDesk, which integrates the computer with the desk, has come into being. It is a deceptively simple table with a computer monitor resting on its marble or wood surface. Keyboard and mouse are cordless, and the computer itself rests beneath the table's surface. But the almost Shakerlike simplicity of the table is driven less by aesthetics than by contemporary security concerns – disks are kept in a lockable drawer, the computer under the lockable desktop; contemporary notions of power, it seems, rest as much on invulnerability as on information. I can't think of a desk that comes closer to being a trustworthy titanic quadruped, yet I hardly think this is what Neruda had in mind.

As important as changeability and security is interactivity. I have seen a proposal for a desk that accommodates both a computer and exercise equipment, and certainly the desks in *Workspheres* reflect an inclination to do other things while working. There is every reason to believe that the designers featured in the show were tuned into curator Paola Antonelli's premise that the changing nature of the workplace was generating some new ideas, but one is also inclined to believe that these designers were influenced by twentieth-century writers who have been vastly inventive in what could only questionably be called their work habits.

Winston Churchill, who did all his dictation from his bed, or Truman Capote, who, it is said, was happiest writing when lying down, might have been the spiritual collaborators on Hella Jongerius's Soft Desk, which is disguised as a bed; its touch pillows function as keyboards, speakers, and floor cushions and are made from gel embedded with computer sensors. And Jack Kerouac might have contributed to the idea of "nomadic work," or at least have invigorated the prose around this phenomenon, which considered not so much the physical furniture of the workplace but "a fictitious system of pills that embody psychological, physical, and personal concepts" of the workplaces and tools "designed for people on the go." One can only wish that Kerouac, a man on the go himself, could have been offered the opportunity to test drive another of the exhibition's "workstations," an SUV designed to serve as a mobile office, outfitted with a global communications system, twelve-camera video system, and digital maps. Particularly had he availed himself of a fictitious system of pills, his wordplay with something called a "global positioning device" can only be imagined. But, of course, no such collaborations or

63

test drives ever occurred; these writers and designers are simply citizens of the same age.

The exhibition was based on the premise of the changing nature of work in the digital era, promoting the interactive desk as a departure from the traditional desk, which was not so much dumb as it was conceptually inert. But while gels embedded with electronic sensors and overhead video screens may certainly bring new materials to the desk, there is nothing necessarily new about desks that perform multiple tasks or even about interactive desks. A common piece of furniture in Georgian England was a man's dressing table with an upright hinged mirror in the center. The table had the same simplicity of the PowerDesk, though the meaning of power here rested on different assumptions. The mirror could be folded down to create a flat surface, transforming the dressing table into a desk and in the process suggesting that the distance between self-reflection and the consideration of world affairs could be efficiently negotiated by a simple hinge attached to a mirror.

In *Mechanization Takes Command*, Sigfried Giedion documented assorted early devices for the reading of multiple texts, none of which would have been out of place in the MoMA exhibition. Scholars of all ages, it seems, favor interactive desks. In *Mechanization*, then, is a rotating reading desk used by the secular scholars of the late fifteenth century, who also had some familiarity with the idea that knowledge is power. The age of humanism, Giedion explained, generated a "growing interest in the bible, the ancient authors, and comparisons of texts." The desk was designed around a central column, much like a lazy susan, with a surface that could revolve at the touch of the reader's hand. Giedion also documented a rotary reading desk designed in 1588 by Italian engineer Agostino Ramelli. Inspired by hydraulics, Ramelli constructed a desk that operated something like a waterwheel, allowing a variety of different texts to be placed on small, tilted platforms, which could be rotated at will by the scholar. Both of these complex mechanical systems allowed the scholars, essentially, to study while they were, well, studying. A scholar does not want to cook or lie down or drive or do anything else while reading; he wants to *read* while he is reading.

Tom Newhouse is a product designer who has spent decades observing how people work and how new technology shapes the workplace. It is his belief that for all the currency given to the idea that the desk is changing to accommodate new technologies, the only real significant innovation is in human engineering – that is, an awareness of how people work. The innovation is in adjustability, he suggests – in desks that can be easily moved themselves and that are adjustable in height and tilt, keyboard trays that can be positioned precisely, a monitor that can be oriented in a variety of different ways. The beauty of the flat screen, he says, is that it is lightweight and small and is thus easier to move and adjust. His observations are echoed by Tung Chiang, a young designer who, in designing a desk, imagined himself to be Galileo. In a presentation of his work in New York City in 2003, he stated, "[Keeping] in mind that the world is not a flat place, I was able to challenge myself to design a more economical table." His table is "full of highs and lows": a monitor at eye level and a keyboard positioned at a lower level, so that the elbow is at a comfortable ninety-degree angle. Elsewhere on the office landscape are modular furniture systems based on spherical joints, tubes, and panels that allow for easy and quick reconfiguration.

Perhaps because of the constant adjusting and readjusting that takes place at one's desk, when Newhouse considers the overall design of an office, he says that "it's healthy to look at infinity." He is talking about ergonomics, about how looking from the monitor to the horizon relieves eyestrain, but he is also saying something simpler about the very nature of working at a desk:

people like to look out the window; a view of the horizon can be sustaining. This is a good way to talk about design – part science, part poetry, part plain sense. Newhouse's words were echoed by New York designer Hani Rashid when he told *Metropolis* magazine in 2002: "In the United States we can build office cubicles a hundred and fifty feet from the window, with no fresh air. In Europe, there's a sixty-foot limit. This code is in place as part of a culture of aesthetics, design, and what it is to be human."

Although Rashid's words might seem to point to a progressive layout for office space, they, in fact, evoke words written a hundred and fifty years ago by Henry David Thoreau, who had his own thoughts about knowledge economy and workspheres. While cleaning his cabin, Thoreau moved his table outdoors, and had this to say in *Walden*:

> ...and my three-legged table, from which I did not remove the books and pen and ink, standing amid the pine and hickories. They seemed glad to get out themselves, and as if unwilling to be brought in. I was sometimes tempted to stretch an awning over them and take my seat there. It was worth the while to see the sun shine on these things, and hear the free wind blow on them; so much more interesting most familiar objects look out of doors than in the house. A bird sits on the next bough, life-everlasting grows under the table, and blackberry vines run round its legs; pine cones, chestnut burs and strawberry leaves are strewn about. It looked as if this was the way these forms came to be transferred to our furniture, to tables, chairs, and bedsteads, – because they once stood in their midst.

Apart from reminding us of our great good fortune that Thoreau never had a laptop, this description of his work table is probably about as good an evocation of a power desk as you are ever going to find. Or, inasmuch as it involves an imagined awning, blackberry vines, and proximity to the everlasting, perhaps this qualifies as an entire office system. And though certainly more eloquent, Thoreau is saying something about an office without boundaries that ads for wireless technology advocated a century and a half later with glossy photographs featuring contemporary information-workers sitting in grassy meadows or on lakeside piers with their iBooks and styluses: a good desk in a good place situates us in something larger than the work at hand. What matters isn't that you can work anywhere, anytime; it is the persistent hope and belief that how you work can, at times, connect you with a larger realm of purpose.

Which is probably what all these alternative desks are really after in the first place. For all their assertive changeability or interactivity, maybe these are just codes words for that window onto infinity. Perhaps Thoreau's imagined awning was the nineteenth-century equivalent of Fukasawa's sky screen; maybe a car as a desk simply wants to drive forever, and a bed as a desk just offers a continuum of dreams. The idea that people merely want to look out the window considers the premise that sitting at a desk is, if not an actual quest for the infinite, then at least an effort to do something, say something, imagine something of genuine enduring value; what we all are after in some way or another in our work is the everlasting growing under, on top of, or around the table.

Infinity can be pursued through any doorway, and a power desk can come in any shape or form. And I think of Neruda's trustworthy, titanic quadrupeds, of a sixteenth-century reading desk that can rotate into eternity, an eighteenth-century scholar with a mirror that becomes a desk and a desk that becomes a mirror, of Fukasawa's information worker glancing upward to the sky screen, and of my son sitting at his first desk in the beach house and looking at the sea.

The MAILBOX

The traffic on our country road moves at a fairly steady speed, so it wasn't until they were seven or eight that my sons were allowed to collect the mail. When they were finally permitted to make the short walk from the porch to the mailbox, it was a passage that represented not only their growing recognition of the complexities of the world outside the door, but also a journey toward identity. The first time they encountered a card in the mailbox addressed to them, they understood that the mailbox was a place where news of the world arrived, and better yet, it was news that arrived with their names on it.

Certainly, most of the mailboxes in my neighborhood live up to this task. While many of them are standard sheet-metal or plastic, others make for a catalogue of decorative sensibilities. Architecture has a foothold here – one is in the shape of a small red barn, another is sheathed in weathered clapboards, others have shingled roofs. Some have simply been painted in cheery colors – forest green, rose pink, purple. One has a small, faux folk painting, a Holstein grazing in a pasture. Still others have incorporated objects that speak to the rural past, propped up in milking cans or set on an old, wrought-iron sewing stand. But even those mailboxes that have no aspiration to the decorative bear some imprint of their owner, whether through plain workaday efficiency or sheer, unapologetic neglect. One of my favorites is a twisted, rusty metal box, askew on an old pole; its varying hues of rust, evocative dents, and precarious tilt are all testimony that the house down the driveway could have been Richard Serra's childhood home.

Cal Swann is an English graphic designer and professor who has documented the mailboxes of rural Australia. There, he notes, the form and decoration of boxes have become their own kind of folk art. Recycled milk churns, oil drums, and petrol cans can all easily have mail slots cut into them, and all provide a secure, waterproof depository. In his visual catalogue of such

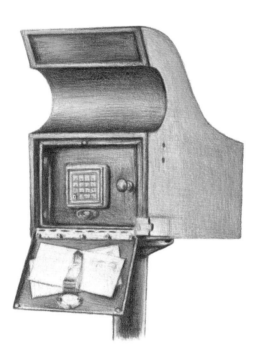

boxes published by Pentagram Design in London under the title *Nifty Places: The Australian Rural Mailbox*, one finds an oil drum reinvented as a pink pig, another as a galvanized silver elephant. Some are pure whimsy, others more steadfastly industrial. Then there are the old refrigerators that have found a second life as mail and package drop boxes. All are examples not only of rural ingenuity, but also of the persistent decorative sensibility that invariably attends it in an unexpected landscape. "Whatever the original motive may have been for the early settler to recycle a ready-to-hand artifact in the performance of mailbox duty," Swann writes, "a cult has developed throughout Australia for the bizarre and often outrageous objects as mailboxes. It now has less to do with practicalities and recyling; competition is rife and efforts to make mailboxes that demonstrate individuality and humour are found in many localities."

While such ingenious folk art may be less prevalent across rural America, it does appear sporadically. Not surprisingly, in view of the fact that design innovation so often emerges from need, some of my neighbors have fashioned larger, more substantial and secure drop boxes for express mail. Consider the high-concept FedEx box constructed by sculptor and furniture maker John Scofield. His lavish attention to the FedEx box extends to the prose he uses to describe the receptacle he has installed outside the door of his Connecticut home: it is, he says, "the holy tabernacle of modern secular life, the vessel keeping body and soul together better than the Noah ever did, the safe refuge for papers of transit that govern one's existential, spiritual, nay, fiscal existence – our FedEx box!"

The aesthetic origins of the box look to both a chicken house ("An assortment of out-house inspired appendages affording a pitiful degree of comfort and usefulness") and Roman architecture ("A temple of mail which mighty Hadrian would have installed on the Pantheon doors if he had expected any serious outstanding invoices"). In fact, the four-light barn sash for a door, traditionally used for windows in barns and unheated outbuildings, was outfitted with brass hinges and latch, and the roof was shingled with cedar shakes and built at a forty-five-degree pitch to discourage build-up of dirt or precipitation and flashed with copper to prevent rain from getting packages wet. An open-air attic is display space for beach stones, bones, other picturesque debris he and his family may pick up. For all its ingenuity, however, Scofield's FedEx deliverer had only qualified appreciation for the box, relegating it to the "second nicest box I've seen." Another customer, it turns out, had constructed a box inside of which was kept an electric coffeepot and mug. Even in this era of urgent, accelerated mail delivery, it seems there remains a time and place for the courtesies of human exchange.

Clearly, the mailbox, a little roadside cabinet printed with one's name, or address, sometimes both, is about personal identity. Inside it is the information, essential and nonessential, that defines one's life; not only the bills and bank statements, but also flyers, advertisements, the series of unsolicited printed matter that in some peripheral way nevertheless helps compose one's identity. Small wonder, then, that the mailbox invites assault. In rural areas, it is a rite of spring for teenagers to cruise by in cars with baseball bats and knock down mailboxes. The kids know it isn't simple vandalism; rather, the trashed mailbox is an assault on personal identity. They know about the symbolic value of the mailbox.

A friend of mine was once a caretaker for an old farmhouse in Vermont that had been the home of Robert Frost. She found Frost's old, gray, sheet-metal mailbox still intact, set at an angle on a post that had settled comfortably into the ground after so many years, his signature hand-painted in black, its sides riddled with buckshot. I haven't any idea whether the latter came from the town critics or just some locals who may have taken issue with Frost's idea of what makes a good neighbor, but whether they read poetry or not, they knew a good symbol when they saw one and recognized that a shot at a mailbox was an assault on the poet's very identity. Indeed, when Frost's farm was bought years later by a Hollywood producer, the old box was replaced with a generic black box. Instead of putting his own name on it, the producer had the name of the farm elegantly stenciled there. You could argue about which of these two men had the greater celebrity, or whose work spoke to his times with more accuracy and resonance. But while the new lettering was certainly not something you would call identity theft, the producer's decision to have the name of his property rather than his own name appear there reflects a distinct dilution of personal identity, not to mention a strictly contemporary menu of choices we seem to consider when deciding how to identify ourselves.

Considering that, it seems almost odd that these boxes are used so rarely for political statement. The U.S. Postal Service prohibits advertising on mailboxes (nor may the post "represent effigies or caricatures that tend to disparage or ridicule any person"), yet you might think the combination of personal identity and public view that the mailbox embodies would make it an efficient forum for political expression. But the closest mailboxes seem to come to that is with decorative stars and stripes or decals of the American flag. It seems strange that in this time when anything and everything can be used as a platform to spell out one's personal convictions, the mailbox has resisted, as though there is some implicit consensus that this would be inappropriate. If there *is* any consistent new image or appendage, it is the security plaque that is affixed to so many mailboxes today. Without so much as a passing glance to the decorative traditions of the mailbox, these square, round, or diamond-shaped signs are strictly informational, naming the security company monitoring the residence. Delivering a terse message of electronic surveillance, such plaques, as ubiquitous as they may be, are out of synch with the spirit of the box itself.

As security becomes a greater factor in how we assess domestic comfort, it comes as no surprise to see the mailbox reinvented. And considering that we live in the age of communication, it follows that this traditional conduit for information has undergone a few changes of its own. In recent years in suburban areas, traditional mailboxes have been replaced by brick constructions into which the mailbox is inserted. Impervious to baseball bats and other forms of garden-variety vandalism, these postal fortifications speak to recently developed concerns about mail security.

But for some, brick enclosures remain insufficient; more extreme measures are needed. Consider, for example, a mailbox offered by a company called Mail Theft Solutions. Once mailboxes were simply associated with their owners' names. The "Defender," trademarked as the "Anti-Theft/Anti-Vandal Curb Vault," is one with a name and identity of its own. Costing roughly $800, the 131-pound steel box is the Darth Vader of mailboxes, constructed of welded plate-steel that its manufacturers pledge "resists baseball bats, bullets, and pipe bombs" and is "the ultimate defense against rising mail and identity theft." Shaped as a substantial pedestal, the box is equipped with ingoing and outgoing shelves. A secured tray on the floor of the unit catches all the mail placed on the incoming shelf, where it then waits to be retrieved from an access door in the rear that is outfitted with a deadbolt. As one might expect, the Defender catalogues "the facts about Identity Theft" in its marketing, citing to prospective buyers that such theft is "the fastest growing crime in America affecting approximately 900,000 new victims each year."

Then there is the Secure Mail Vault, a white, powder-coated, galvanized steel box with a gracefully curved mail-slot top in the shape of a wave that looks like something Frank Gehry might have thought of but has, in fact, been so configured to prevent a burglar from reaching an arm into the vault. The vault is outfitted with an outer compartment for outgoing mail, and an inner compartment to hold incoming mail that can only be accessed by a keypad, with which its owner must validate his or her identity. "Protect you and your loved ones from the dangers that linger in and around your home," advises the promotional copy. And I recall Robert Frost's bullet-riddled mailbox and wonder whether he would have purchased something called the Secure Mail Vault and what he might have had to say about the dangers that linger in and around the home, not to mention which words he might have found for the notion of identity theft.

Perhaps it is only fitting that the electronic age has reshaped the mailbox. Identity theft and mail fraud may be legitimate security concerns. And at a time when information of all kinds comes to us in a steady and constant electronic stream, the daily walk to the mailbox has become an outdated ritual. Yet the image of the mailbox seems to have its own tenacity. In 2004, in an effort to encourage people who lived in remote rural areas to vote using the absentee ballot, the Democratic Party in Arizona sent out a flier picturing a worn mailbox propped up on a pole along a dusty road. Without keypad or lock, and without so much as a nod to the idea of identity theft, which one might think would be at issue in voting, the image simply conveyed a comforting familiarity. "Now it's easier to vote than ever before," read the copy.

It only makes sense that Mail Boxes Etc. still relies on that traditional nomenclature. Established in 1980 for shipping, postal, and business services independent of the U.S. Postal Service, twenty years later, the company had over four thousand locations and had merged with UPS to become the world's largest franchiser of such services. The company's stated core values include "caring, honesty, fairness, integrity, trust, respect, commitment, accounta-bility" – a laundry list of traditional community values. As well as the shape of the box and its little red flag, then, even the word "mailbox" seems to have an iconic resonance, exactly because it implies an old-time trust, a nostalgic view of communality, an era when, despite the occasional threat of bullets and baseball bats, neighbors, passersby, and the mailman were all members of a community that conferred its own brand of security. For all the printed matter that goes in and out of these boxes day after day, their most resonant message may simply be one of implicit trust.

As I drive on the country roads in my county, I can't help but notice that there is not yet any sign of the Defender or the Secure Mail Vault. One neighbor recently repainted his blue mailbox, and another has hung hers on a trellis and planted it with a vine of clematis. None of these has locks or padlocks or electronic keypads, though they may come yet. And I know that somewhere out there, someone is brewing a pot of coffee in expectation of a delivery from FedEx. From time to time, even the sterile aluminum or stainless-steel cluster mailboxes used in residential apartment complexes acquire a pitched roof or fluted support columns, decorative gestures that seem an effort to asso-ciate them with their roadside predecessors. And for all their instant messaging, my sons continue to take pleasure in checking the mailbox. This could be because, for the moment at least, it puts our vigilance at bay. A vestige of rural neighborliness, it offers a different kind of security.

It seems almost astonishing that in the day of electronic defense systems – of digital passwords, laser home surveillance and keypad auto access – we continue to use the roadside mailbox at all. But maybe that is the point exactly. At a time when all manner of communications equipment has been reformed and reinvented entirely, this small roadside cabinet, at once conspicuous and assailable, has an undeniable tenacity in continuing to be just as it always has been. Whether a simple green plastic box, a small red barn, an old oil drum, or even buckshot-riddled sheet-metal, it is a piece of communications equipment that knows its own vulnerability, that accepts all the frailties implicit in human exchange.

the VEGETABLE PEELER

When I was very young, my family and I lived in Japan, and a woman who lived with us then and took care of my sister and me introduced us to the sorcery of common objects. If you are longing to have a visitor leave your house, Masako Ohara told us, you need only a broom and a handkerchief. These two common objects are the accessories of exit. Tie the handkerchief around your head and sweep the dust out of the kitchen and onto the porch, and within minutes your guests will excuse themselves and be gone. Such is the resonance and power of simple kitchen objects.

It is impossible not to think of her when I look around my kitchen, because it is a room full of things that have acquired a significance and power of their own. I think of her, for example, if I happen to be putting away food in a small plastic container designed by Morison Cousins. While it is entirely possible that Cousins was not inspired by Japanese folklore, surely he, too, understood the resonance of common kitchen objects. Why else would he have designed a colander that looked like the starship *Enterprise*? And when he designed Tupperware, it was with some understanding that ordinary little plastic bowls could become not simply indispensable food containers, but the agents of social gatherings as well. In 1951, the Tupperware Company hired Brownie Wise, a middle-aged, divorced mother who had suggested that throwing little parties might be a way to persuade women to buy the containers, and the company has since stated that a Tupperware party begins somewhere in the world every two seconds. Surely that is a fact worth contemplating when you are putting away a plate of leftover food. But this is how it is with the ordinary kitchen objects – a stack of small plastic containers can make your house a place where people might gather, while a broom and a handkerchief are all that is needed to make them leave.

More than any other room in the house, the kitchen is full of utensils that have a power that extends well beyond their function. Probably, it begins with the food. A friend of mine told me of a simple picnic she once arranged. There were assorted sandwiches and salads and cool drinks, and she had placed these all on a picnic table in her garden, under an old oak tree. But the charm of the outing was short-lived; the random chemistry of ingredients had an unexpected toxicity. Apparently, the fumes from the oak tree, the resin from the German wine, and the pistachios in the ice cream created a constellation of poison; headaches, fatigue, and nausea soon followed. Malice converged unexpectedly. Such are the unpredictable alliances and misalliances that can occur in the kitchen and that can only begin with the food.

They extend from there, quite naturally, to objects. It doesn't hurt that many of us were raised watching the dancing spoons and talking teapots of Looney Tunes and Merry Melody cartoons, and have little trouble entering into a verbal, if not emotional, relationship with kitchenware. In his book, *A Box of Matches*, the author Nicholson Baker imagines "the adventures of a cellulose kitchen sponge that somehow in the manufacturing is made with a bit of real sea sponge in it, giving it sentient powers. It lives by the sink but it has yearnings for the deep sea; it thirsts for the rocky crannies and the briny tang." Surely there are few of us who would have trouble accepting a sponge with a collective memory; while manufacturers of the information age seem to be keeping themselves busy trying to imagine ways in which appliances can be given intelligence with silicon chips, they might do better to simply recognize what most of us already know – in the kitchen, at least, what we are really after are objects and appliances equipped with emotional intelligence.

Advertisers, of course, know this. Their line of work celebrates our willingness, our compulsion to attach a voice, a face, a persona to common objects, sometimes universal, sometimes specific to a moment in history, and for generations they have tapped into the resonance of kitchen objects, readily attributing all kinds of characteristics and qualities to inanimate objects in their quest to establish brand identity. What are called "brand mascots" have been a way of life and commerce in America ever since the plate ran away with the spoon, and their various personas range from the charming to the preposterous. In its ongoing quest to characterize small squares of absorbent paper with floral patterning as heroic, Georgia-Pacific, the parent company of Brawny paper towels, found it necessary in 2002 to reimagine the Brawny Man for the post-9/11 kitchen and reconfigured the traditional mustachioed lumberjack into a "tough yet sensitive" fireman. In 2003, an $85 million national advertising campaign introduced the Oven Mitt as the spokesthing for the fast-food chain Arby's; speaking in the voice of Tom Arnold, the talking mitten was imagined to be a persuasive vehicle with which to deliver the message about the nutritional value of traditionally roasted food.

Predictably, advertisers in recent years often relinquished talking objects for the imagined sophistication conferred by celebrity endorsements. But as Morison Cousins or Brownie Wise or anyone who has ever chatted with a kitchen appliance could probably have told them, talking objects convey considerably more emotional depth and resonance than many celebrities. A variety of familiar mascots are now simply asked to keep up with the times. The Energizer Bunny is no longer constructed of pink fluff, but high-tech silver. Even Mr. Peanut, first imagined by a Virginia schoolboy in 1916 who won a contest with his drawing of a "little peanut person," and who has persevered for generations as the persona for the salted snacks division at Kraft Foods, has recently learned to dance and play basketball in an effort to associate the snack with a healthier lifestyle. In the language of design and ergonomics, the word "transgenerational" refers to something that can be used easily by anyone from the age of nine to ninety, but there is something about a bunny learning to dress itself in techno-textiles or an eighty-eight-year-old peanut learning to play basketball that captures the sense of the word more fully.

Even the simple names of kitchen objects often acquire a meaning that goes well beyond their domestic function. Think of the Cuisinart, introduced in 1973 at a Chicago housewares exhibition. With a name that managed to express its own mix of art, Gallic culinary sophistication, and ordinary American domesticity, it enabled a generation of suburban housewives to become proficient chefs; with such transformative powers, it is small wonder that it became a little icon of its time. Teflon-coated cookware developed at DuPont not only encouraged us to cook without butter, but also became a metaphor for all sorts of corporate and political invulnerability.

In considering the line-up of culturally meaningful icons, it occurs to me that the Good Grips kitchen tools speak even more clearly to these times. We are willing to believe that kitchen accessories can convey innumerable human characteristics, from humor to masculinity to congeniality, but Sam Farber, the founder of OXO International, asked why they could not also speak to kindness. With a degree in economics, Farber had started Copco in 1960, producing a line of steel and enamel kitchenware. In 1982, he retired to the south of France. But watching his wife, who was afflicted with arthritis, trying to grip a potato peeler, he wondered what could be done to help her and the twenty million other Americans suffering from the disease. Farber teamed up with Smart Design in New York City to reconfigure, among other things, the generic metal vegetable peeler first introduced in the early 1900s, which with its narrow steel handle can be sharp, awkward to hold, and prone to rusting.

The designers researched the way people utilize such kitchen devices, cataloguing their variety of gestures and motions into twisting and turning; pushing and pulling; and squeezing.

Working with models, they realized that most kitchen tools demand a combination of these motions. For the peeler, then, the handle was enlarged to improve leverage and reshaped as an oval so that it wouldn't rotate while it was being used. The metal handle was replaced with Santoprene, a non-slip, polypropylene plastic and rubber material originally used for dishwasher gaskets. The designers also shaped the handle to include flexible fins that can conform to individual grip, preventing the peeler from slipping. An oversize hole made it easy to hang, even for people who might be visually impaired.

For all the qualities kitchen utensils seem to convey, here was one that not only reflected efficiency and safety, but in recognizing the need of pretty much anyone who would ever use it also embodied even more basic ideas about consideration, empathy, and comfort. And while the new vegetable peeler was certainly easier for people to use, whether they had arthritis or simply the limited strength and mobility of the elderly, its new shape had a kind of funky elegance that made it appealing to everyone; its congruence of high design and utility managed to establish a new standard for kitchen utensils.

The line of OXO Good Grips kitchen accessories that followed is widely available today and includes everything from pizza slicers, graters, and mincers to corers, spatulas, and whisks. All of them, the company notes, are designed "to make everyday living easier." This is a simple statement, and one that probably reflects the simple origins of the vegetable peeler. It is probably worth remembering that despite the months of human engineering research, despite the facts that this is a "transgenerational product," that production costs in Japan necessitated finding a new manufacturer in Taiwan, that this line of utensils is now sold worldwide in more than thirty countries, and that the vegetable peeler now resides in the permanent collections of The Museum of Modern

Art and the Cooper-Hewitt, National Design Museum, it began with nothing more basic than a man wanting to help his wife. Call it a brand mascot for human empathy.

I wouldn't know what criteria were used to deem the vegetable peeler a worthy addition to these museum collections, but I do know that objects favored by museum curators must often transcend their function. And a vegetable peeler, I learned, is capable of doing this. It was a lesson brought home to me when my sons were in middle school. When Noel was in sixth grade, his curriculum included a class called Home Careers. It is what was once called Home Economics, where students learned how to cook and sew and do laundry, that is, where they were familiarized with simple domestic chores. When he came home from school, I asked what sorts of things they had done in class. And he said with all the exasperation and world weariness that only a twelve-year-old can muster, "I learned how to apologize."

My first reaction was a mix of curiosity and horror. But as I thought about it more, it seemed to make its own kind of sense. It occurred to me that there is not nearly enough apologizing in this world. How different life might have been for Trent Lott, Pete Rose, Henry Kissinger, Cardinal Law, and a seemingly endless list of other public figures if only they had taken Home Careers with sixth-grade teachers who could have familiarized them with the value of a decent apology, spoken sincerely at the right moment. Imagine a world where apologizing is taught, if not as a classroom skill, at least as one of life's necessary lessons. Imagine a corporate CEO knowing how to say, "I am just so sorry I plundered your life savings and robbed you blind with my yearly bonus and stock options and my accountant cooking the books. I'll live with regret till the end of my days." Or imagine, possibly, a manufacturer capable of the words, "I am just so very sorry that the stroller," or tires or cars or car seats or whatever the product might be, "was so incredibly poorly designed and manufactured that it resulted in grave personal injury. I will live with shame for the rest of my life." When an apology is publicly stated, it can get world attention: when Richard Clarke apologized to the 9/11 commission in the spring of 2004 for his part in the failure to anticipate the terrorist attacks, his words were met with stunned gratitude from families.

Truly imagining that Noel was on the threshold of a new epoch in human social relations, I asked him, "So, how do you apologize?" He replied, in a singsong voice accompanied by some eye rolling, "I'm sorry I hurt your feelings." It may not have been much, but it was a start. And while he clearly had not mastered the art of apology, the idea stayed with me, and it struck me that there could be a genuine association between the domestic arts and ordinary behavior; and that this sixth-grade teacher whose curriculum had included the seemingly outlandish idea that there might be a connection between learning to prepare food and learning how to apologize was onto something; and that possibly ordinary household tasks can dovetail neatly with larger acts of human decency.

Which brings me back to the vegetable peeler. Not so long after this, I saw an article in the newspaper about an incident that occurred in Berlin in 2002 involving an object called a *Gurkenhobel*, not unlike a vegetable peeler, but actually a kind of cucumber slicer, made out of plastic with a steel blade, that is used in Germany. It turns out that the West Germans produce efficient, durable cucumber slicers, so on a recent visit there a family of East Germans had bought one for twenty-five euros. Once they got home, however, they were less pleased with how it worked, so they wrote an angry letter to the West German manufacturer outlining its failings.

More irritated correspondence ensued, a four-star chef tested the product, the exchange was written about in the press, and it ultimately became an international incident with political and cultural overtones, with the West German manufacturer writing,

"I recommend you go back to using your former East German scrap metal because you cannot cope with the demands of a valuable high-quality product." In the end, of course, everyone involved in the incident apologized. It turned out the West German manufacturer was, in fact, originally from East Germany, and the mayor of his town apologized to everyone in both the east and west. The manufacturer weighed in, stating, "I'd like to apologize to the people in the East. It was not my intention to attack them." Everyone just said they were sorry.

This episode, it seems to me, reflects some basic truth about design. Because it does seem apparent that the shape of the things we use, these ordinary kitchen utensils like potato peelers and cucumber slicers, can engender not only better living, but also better human behavior. When you think of it this way, it makes all the sense in the world that the OXO vegetable peeler came into being because a man wanted to help his wife; the result was an act of kindness disguised as a kitchen accessory. Possibly this is what the OXO company means when it talks about making everyday living easier, and maybe it is what my son's sixth-grade teacher had in mind as well.

I wonder if it is because these are hand-held objects that we use to prepare food – and by extension to nurture and sustain ourselves – that we are willing to attribute near-human qualities to them. It seems inevitable that such items, almost in spite of themselves, are instruments not simply of food preparation, but of human behavior, coordinates that can help us calibrate our place in human relations. Whether they are talking mittens that try to improve the way you eat, Tupperware parties, or a Japanese woman trying graciously to ask visitors to leave and turning to her handkerchief and a broom, any and all of these can be the small agents of human decency.

The TELEPHONE

When I was growing up, news of the world arrived through a kitchen wall phone, a household staple of the sixties. Designed to complement the wall-hung cabinets, it was an artifact of domestic efficiency that acknowledged both the conversational and the housekeeping needs of the homemaker. But in my house, the white wall phone served another purpose as well. In those days before answering machines, it sounded out the uncertain rituals of human communication. News arrived irregularly. You got the call or you didn't, and uncertainty attached itself to the way people conveyed their thoughts. Like the box itself, suspended on the wall above the red Formica counter, its clearest message was that words come unpredictably and out of nowhere.

Such subtle lessons about human communication now seem outdated. Today some 20 percent of American households have multiple phone lines that allow family members to make and receive calls and faxes without getting in each others' way, transmit data, browse the Web, provide online services to home security; by 2003, 67 percent of American households had wireless phones at home as well. With multiple lines, cell phones, voice mail, e-mail, and fax machines all allowing for a steady flow of information in and out of our lives, being connected has become essential to our idea of domestic comfort.

It's hard to complain about telecommunications that make us more available to one another, that help parents keep close track of their kids or employees in constant contact with their offices. All the same, it's with a sense of loss that I realize my sons rarely run to answer a ringing phone with much conviction or urgency. They know a machine is there to take the call. Nor with their instant messaging will they ever wonder who has been trying to reach them, or ruminate about missed calls and lost chances. That insignificant and ordinary information can take on a sense of urgency may be one of life's smaller lessons, but they are unlikely to learn it from the silence of a telephone that has just stopped ringing. They may yet discover that spontaneity and serendipity are an essential part of human communication, but phone calls just caught or missed won't be what tells them that.

It was the absolute unreliability of the standard phone from the sixties and seventies that transformed it from a simple domestic appliance to an inevitable accomplice in our emotional lives. As an agent for the human voice, what else could it be? Emotional collusion with the telephone was the norm; how could you not despise a telephone that didn't record calls missed in your absence? Yet when an anticipated call finally arrived, it could elicit satisfaction, bliss. Rage, resentment, rapture were all feelings that might be legitimately harbored toward a telephone. And that emotional load was conveyed by the physical presence of the telephone. The older generation of phones were not simply able conspirators. They looked the part. If these phones were not entirely reliable, certainly they appeared that way. The black table phone, the Model 500 designed by Henry Dreyfuss in 1949 for Bell Telephone Laboratories, became a classic. Dreyfuss had consulted with Bell Laboratories since the early thirties to reimagine the form of the telephone, and with epic understatement, considering what was to come, remarked that "they wanted a little art to wrap the phone in."

While it may not have been art exactly, the black Bakelite rotary-dial receiver that was used for the next twenty years expressed an undeniable sense of authority. Its stocky, substantial form clearly suggested that if news of the world was to come, it should arrive through a solid and capable conveyance such as this. The design fully recognized that this appliance occupied a highly charged and volatile emotional environment: the handset for the unit was shaped so as to withstand being slammed back into its cradle. How the phone was marketed also reflected gender biases of the time. As writer and curator Ellen Lupton writes in her book

*Mechanical Brides*, "For the male executive, the phone [was] a vehicle of progress comparable to the train, plane, or automobile," while for women, it was a social tool: "female phone conversation serve[d] nurturing functions."

Using the 500 as a kind of consumer baseline, Bell and Dreyfuss brought to the market a whole host of other telephones – wall phones, multiline phones, TouchTone phones. The 500 was available only in black for the first few years, but colored models were quickly made available. Consider the historic resonance during the cold war of the red telephone, on which vital communication between the superpowers rested. "This is one of the most important telephones in the free world," ad copy for Bell solemnly intoned.

But with all its assumptions about femininity, the pink Princess telephone introduced by AT&T in 1959 to the newly discovered market of teenage girls is probably the most memorable communication icon of that time. Designed as a bedside phone, it both reflected and nurtured the new idea of privacy for teenagers. Likewise, its shape and light weight allowed it to be easily carried in one hand. "Ads for the new oval shaped phone emphasized is small size and light-up dial. Petite, horizontal, and cast in decorator colors, the Princess phone suggests a reclining nude," writes Lupton. For sheer expressiveness, though, these versions pale in my mind beside a phone I once bought in the late eighties in a home appliance megamart. The phone was in the form a shiny red high-heeled shoe – the heel of the shoe held the listening device, while one spoke into the gracefully pointed toe. I gave the phone to a friend who believed deeply in the transformative power of shoes, red ones especially. This was not footwear for a trip down the information highway; it was an implement of courage and imagination. The first call my friend made on it was to break up with her boyfriend.

The phones my family use today bear no resemblance to their expressive predecessors. The taupe, neutral, squared-off form of the cordless phone in my kitchen attests to a bland, interior life, and the cell phones my sons use have a similar banality; there is no doubt they are gender neutral. Their ever diminishing size suggests an equally sparse emotional content, and for all their presumed efficiency, these are phones you can't slam down. The styling of most mobile phones seems interchangeable, and their color and form hold little interest, unless, possibly, it is in the obsessive reconfiguration of their keypads or their ever diminishing size. Certainly manufacturers make a token effort toward individualized expressiveness in offering customized faceplates. Consider the Nokia phone with interchangeable covers marketed as "ranging from bright crimson red to aqua blue with glow-in-the-dark butterflies [that] enable customers to instantly customize their phone to match their mood, attitude, or outfit." As one marketing executive noted, "Customers see their phones as a reflection of themselves." That was no doubt the case of a more high-end phone customized for the wife of a rapper: the Motorola was encrusted with sapphires and diamonds at a cost of $125,000. Still, whether embellished with butterflies or diamonds, these seem remote from what Dreyfuss had in mind when he tallked of wrapping the phone in a little art.

Such miniaturization even accommodates its own hostilities, and I wonder if there is something in this new generation of communications technology that reflects a still more subversive emotional life. Throwaway phones introduced in 2003, purportedly reflecting a sense of carefree, easy living untethered to costly and restrictive contracts, revel in their own breezy malice. Using flexible circuit board technology, the phone is essentially a phone card with dialing capabilities, its variety of prices determined by its increments of talking time. And while it may not be very satisfying to slam down a phone card, it does lend itself to a gesture at once more casual and more hostile – simply being discarded, an act that conveys an implicit message about the dispensability of human exchange. With a certain minimalist elegance, the disposable phone recognizes in its own way that human communications are an emotionally loaded enterprise: throwing the phone in the trash is surely the ultimate and decisive gesture of dismissal.

Dismissal, remove, unavailability, and unresponsiveness are all vital to human exchange; communication itself is established on the option of remaining silent. The white wall phone in my mother's kitchen spelled this out simply and clearly. Sometimes it rang. Sometimes it didn't. The cell phones my family use today fully recognize this tricky, nuanced territory; though they may be unintentional, flaws in wireless technology accommodate the uncertainties of human dialogue with a sophistication infinitely greater than the white kitchen wall phone. It was hard to be disappointed when I discovered that my sons' cell phones have achieved their own brand of unreliablity; it is not so direct or blunt as that of the phone I used as a teenager, but rather operates with its own capriciousness. These are phones that are as expert at separating us as they are in connecting us; they are unpredictable, elusive, complex – which is to say, their qualities are human. Our voices fade without warning; words are cut off, connections abruptly terminated. "I'm losing you" has become common parlance in cell phone exchanges, so much so that those three words recently served as the title for a novel about Hollywood, the film industry, and the frailty of human relations therein. These are phones that know that even when we are talking, sudden disconnection is imminent.

Their diminutive size makes the phones easy to lose, and for days my sons can be without their electronic tether. In addition, while cell phones are generally perceived to service human communication, not answering them and not even checking for messages is often perceived as the ultimate luxury. For all we do as constituents of the communications age to acquire new accessories, we then regain the upper hand by ignoring them. Considering the human taste for contrariness, its not surprising that in the era of exponentially multiplying communications equipment, being unreachable has quickly attained a status all its own.

Besides, most of these phones have other things to do. Just as their size is ever diminishing, their range of functions only increases, and industry spokesmen routinely inform us that phones that serve exclusively as agents of communication will be antiquated within the next few years. And so they accumulate features, one after the other, including memo pads, alarm clocks, to-do lists, calculators, calendars, and rings that resonate with everything from Beethoven to the Beastie Boys; they can download games and ring tones, browse the Web, receive e-mail, and send text messages, while the built-in eye of the camera transmits documents and candid moments. Soon enough, we are told, phones will also be hooked up to Web-cams, enabling their users to check in at home on pets, kids, domestic help, sick family members, or whoever else might benefit from such surveillance. Other technology connects phones to wired homes to operate appliances remotely. A French telecommunications company is working on a model that tracks locations so that parents will know where their kids are just by picking up the phone, while in South Korea, LG has recently introduced a phone aimed at elderly users that allows them to check their glucose levels. So it occurs to me that the most resonant message of the latest generation of tirelessly multitasking phones is something I could have told you when I was a teenager, and that is that words alone are never enough.

We know how these phones have affected social exchange – they make us late, they make us rude, they make us remote from the people we are with because we are chatting with people we are not with. No wonder, then, that in recent years, the nostalgic appeal of vintage phones from the forties, fifties, and sixties has asserted itself, and we see everything from solid black Bakelite desk phones to restored wall phones to Princess phones retrofitted with push buttons. A designer in England has gone so far as to design a prototype attachment with which old-fashioned heavy receivers and their spiral cords can be connected to cell phones. The incongruity of such a phone surely makes for a comic sight, but it also says something about the materiality that human conversation sometimes demands. Still, there is an implicitly inconclusive quality about such an object, with its unsettling sense of dismemberment, and I wonder what Dreyfuss would have to say about a phone that supplied a handset without a cradle in which to slam it.

James Biber is a New York architect who attests to a growing number of clients who use vintage phones. He has one himself. The deep, resonant ring of these old phones is satisfying, and he talks about how "wandering around the kitchen, tethered to that twenty-five-foot mess of phone cord with a big, heavy handset nestled between your shoulder and ear is the only way to have a real conversation. Everything else is just digital." The qualities he mentions about vintage telephones – their sound, their weight, the time they seem to allow for – all have their corollary in the talk itself. Maybe we value these substantial phones because we value exchanges that might be of greater substance, that carry with them time, weight, a beautiful sound. If there is any single message the old phones convey, it is that human exchange is worth more than the hurried, elusive exchanges we make on cell phones, conversations that begin with "How are you?" instead of "Where are you?"

A woman I know tells me that her grandmother had a telephone room – a small, separate room lined with floral wallpaper and furnished with a comfortable armchair chair, a lamp, and a table with the telephone, notepad, pencil, appointment book, and address book – and it was where her grandmother went to sit when she made a call. "Taking the time to talk required a room with a door. You needed to have your privacy," my friend tells me. I don't know that my friend's grandmother had a room of her own, but her telephone apparently did. And I can't help but wonder if such rooms will also make a comeback, as a kind of spatial antidote to the vast placelessness of calls made from cell phones. Nor does it seem especially far-fetched to imagine that a year or so from now, the McMansions that are being built across the country and marketed as the ultimate in gracious living may include, along with the spa baths and home theaters in their vast catalogue of perceived amenities, telephone rooms that offer privacy and solitude.

No sooner do I imagine this than I realize a similar place already exists. Every summer, my family and I spend time at a barn on the side of a hill in Vermont. Part of the beauty of the barn has always been its primitive quality – rudimentary plumbing, wood stoves, and the absence of a phone jack. Being out of touch has always been part of the pleasure of the vacation. Our cell phones, of course, changed all that. Still, cell service in the area is intermittent at best, and it was several years before we discovered that there was a granite rock in the middle of a pasture that was a likely spot for our phones to get a signal. When I first saw one of my sons perched on that little slab of speckled granite in the middle of the field, I couldn't help but be charmed by the realization that his efforts to find a good place to talk situated him in a primitive yet parallel place to that wallpapered room occupied by my friend's grandmother some fifty years ago.

In the end, though, the question remains: Which is the more trustworthy appliance for human exchange? Is the Samsung 310 really preferable to the Bell 500? Which one is the mechanism more able to capture and convey the unpredictable rhythms of communication? A phone that rings with the beat of Brazilian jazz, that can send a silent video clip of ocean waves, that enables me to play Clue while I am waiting for a call, giving equal significance to the questions – Did I get the job? Did Professor Plum use the rope in the dining room? Will you call me back? Or is it a more substantial phone that has weight, a beautiful sound, and conveys time? Is it a phone that itemizes missed calls or one that simply misses calls? Is it one so small that it can be lost without thinking or one with heft and stature that can be thrown across the room with pleasure and satisfaction? Is it one that has a sounds menu or just one with a clear, deep ring? Is it one that has a feature called Whisper Mode that transforms my voice to a soft murmur or one that carries my voice with clarity and inflection?

I imagine the telephone of the future to be one that has somehow managed a convergence of these capabilities – it can take a picture of the mountain I am looking at outside the window as we talk and send it to you, it has a clock, it rings with the music of a Portuguese choir, and when we are finished talking, depending on how the conversation has gone, I can play Tomb Raider. It manages all these tasks, yet it has its own weight and substance. It is Ozone Blue and has a hefty handset that is naturally cradled in the hand and a heavy coiled cord that I will wrap around my finger, my hand, coiling it and uncoiling it as I try to imagine what to say next, and all of these reinforce what I know, which is that talk between people is often a physical enterprise, and when this occurs the telephone is called upon to be a prosthetic device.

But even as I envision this telephone of my dreams, I remain certain that despite all the appliances with which we fill our homes and lives, our connections remain elusive. And I recall those years of being a teenager, listening for the ring of that white kitchen wall phone, my accomplice, my witness to the ephemeral relationship between words and time. Its silence suggested that communications observes its own cycles. I learned then that words come, they go, you never quite know when, and I think of that time as the age of information.

the REFRIGERATOR

Design, we are often told, is about communication. The statement is so general that at times it cannot help but be true, but its generality can also cause it to be wrong. Besides, who ever said communication was always something to aspire to?

The question is raised by the current advent of "smart" machines, the result of hooking up wireless communications technology to domestic appliances. With the help of a modem, such hook-ups allow refrigerators, laundry machines, ovens, and even toasters to communicate with the owner, manufacturer, and, of course, one another. Such appliances, falling into a new category of kitchenware known now as "infopliance," include curiosities like "smart" dishwashers capable of timing the use of hot water with water utilized elsewhere in the house, a washing machine that can be operated by a cell phone, a microwave oven that can go online for cooking instructions, and a toaster that delivers personal greetings.

But it is the refrigerator, it seems, that has the loftiest aspirations, that has set out to be the smartest of all domestic appliances. And that comes as no surprise. While it is commonplace to think of the television as the electronic hearth, it is surely the refrigerator that serves as the family's comfort and control center. It only makes sense that the designers of Refrigerator Freezer Rack (RFR), designed collaboratively by NASA and its European counterpart, ESA, and planned for the International Space Station in 2006, took into consideration the comfort factors of refrigerators; there is probably no place where comfort is more important than outer space. One of its designers, Margarita Bergfeldt, specifically addressed the emotional language of the multi-compartment unit – handles, food trays, and digital interface were all designed for comfort foremost – and she told *I.D.* magazine in 2004, "I wanted to make it a system that's not technical but homey. People shouldn't feel like they're eating an experiment." That one solution for such hominess was a food tray embossed with a Gubbe – a cartoonish, anthropomorphic figure – is another matter.

101

Ordinary, earthbound refrigerators offer comfort more improvisationally. Though not necessarily designed to serve as such, the vast, white, magnetic surface of the traditional refrigerator has become a blank screen across which the minutiae of family life play. With its haphazard arrangement of children's drawings, invitations, postcards, appointment reminders, the place where we keep our food cold has also managed over the years to become a sort of spontaneous domestic communications center. The refrigerator *needs* to be the smartest.

For better or for worse, then, manufacturers and designers have set out to improve upon improvisation. Consider the LG refrigerator, said by its manufacturers to "connect this hub of the household to a vast matrix of information and entertainment available through advanced multimedia-computer technology,... a residential gateway to the home." Aside from keeping food cool, such a refrigerator enables one to go online, shop, check e-mail, download music, watch TV, monitor grocery inventory. The Screenfridge is billed by its manufacturer as "a refrigerator that thinks." Further advancing the idea that the refrigerator is the family command center, it comes equipped with a computer, keyboard, touch screen, speakers, microphone, and even a small video camera, enabling family members to leave video or e-mail messages for one another. When connected to the Internet, the refrigerator can scan bar codes and order groceries, while its electronic tagging system enables it to keep track of when the milk sours or the cheese goes rancid. And in the spirit of hybridization that seems to be essential to so many contemporary objects, there is Whirlpool's Polara Refrigerated Range, which takes its cues from a thermos bottle and makes the most of insulating properties. The unit has a baseline temperature of forty degrees to cool food, but the temperature can be adjusted upward to cook food as well.

At the other end of the spectrum was Audrey, billed as the "first digital home assistant." An online family organizer available in five colors, it set out to document and store the random information that collects on the refrigerator surface with its date-book, address book, and calendar, while its Internet hook-up could access everything from the local weather to the stock market. The small countertop appliance set out to eliminate the visual clutter that accumulates on the refrigerator, cleaning it up, sanitizing its spontaneous chaos and reordering it into a seemingly more accessible format. Still, that it was simply another countertop appliance didn't help its cause; the beauty of the refrigerator as bulletin board has a certain double-duty, improvisational efficiency that appeals to most people far more than the addition of yet another machine requiring its own space. Not surprisingly, Audrey was discontinued soon after its introduction, and its short life probably offers us a lesson about recognizing the value of unplanned secondary uses; there is something in the human imagination that thrills to these renegade functions of an object meant for one thing being utilized for something else altogether.

Still, Audrey points out that we have come a long way since Mr. Coffee set out to anthropomorphize kitchenware. But for all its presumed efficiency, the kitchen as a chat room for appliances can conjure up more disturbing scenarios. Writing in *I.D.* magazine in 2000 about the needs that smart appliances fill in our lives, Peter Hall cited a British entrepreneur who has been marketing an audiotape of background household sounds such as those of a shower, a vacuum cleaner, and a hair dryer. "The intention aims to fill the vacuum of bachelor life," he observes. "Hit the play button and your home hums with the sounds of human company. Smart appliances perform the same function, only for more people and with more tangible results. With the coffeemaker gurgling away in the kitchen as we wake up, we won't wake up feeling so alone." Or will we? It seems more likely that a congregation of these appliances might not evoke the cheerful communality of relaxed family life so much as a sentiment once voiced by Lillian Hellman that "lonely people talking to each other can make each other lonelier."

Because what catches our attention, of course, is the assumption that speech is necessarily an indication of intelligence. You might think we had gotten over this conceit by now. Elsewhere on the cultural landscape, there seems to be some acknowledgment that talk is cheap; political campaigns offer us ample opportunity to recognize the vacuity of both stump speeches and the commentary surrounding them, and in the realm of psychotherapy the value of the talking cure has been largely replaced by pharmaceutical treatments. That compulsive chatter may reflect anxiety, insecurity, or any host of personality disorders more than it does intelligence has not yet been accepted by the manufacturers of domestic appliances, who persist in believing that talk is something to aspire to, that something that speaks is something smart.

Maybe design *is* about communication. But if so, it is probably about the sort of communication that is a more elusive enterprise, one that recognizes that silence and knowing when to keep silent are essential to human exchange. Frigo Design has done this by manufacturing a chalkboard surface that can be affixed to a refrigerator. In some kind of effort to be a smart appliance in the more primitive meaning of the word, the green or black panels attach to the front of the refrigerator to suggest the kitchen is a classroom, though clearly it is a pre-Internet place of learning. Encouraging a more traditional means of exchange, the board can be filled with chalked messages, notes, and reminders. Just as easily, it can be wiped clean, become blank.

Surely that quiet is essential to anything that considers itself smart. And if the refrigerator is the center of family communication, it probably serves most efficiently when it acknowledges dialogue is not simply a series of brisk, efficient commands passed between one appliance and another, but rather a more subtle exchange that recognizes not only silence, but more tentative utterances as well. Anyone who has ever been a member of a family knows this, and any appliance that sets out to help people communicate should probably recognize it, too.

Like our emotional relationships with human beings, those with objects tend to be unpredictable and elusive, and I suspect that research labs working on developing robots with emotional intelligence have a long way to go. While we can hardly deny that some kind of humanity is at work in a refrigerator that anticipates nutritional needs or orders food directly from the grocery store, Banana Yoshimoto's book *Kitchen* offers a view of a more genuinely emotionally responsive refrigerator, one that is a gentle witness to grief. After the death of her grandmother, the character Mikage Sakurai finds that the only place she is able to sleep is curled up next to the huge, stainless steel refrigerator "stocked with enough food to get through a winter" and whose hum "kept me from thinking of my loneliness." And I wonder if it is possible for such an image and such words to figure into the conversation when researchers and designers speak about the emotional appeal of design.

If our refrigerators are bound to be equipped with intelligence, there is another means of expression, though more primitive than modem-enabled dialogue, that is also particular to the refrigerator and that does just that. I am thinking of the subtle, more nuanced exchanges enabled by refrigerator magnets. A literary phenomenon of the nineties, these small, white magnetic tiles each printed with a single word in black were an admittedly unassuming collection that nonetheless had the grander ambition of taking "poetry out of the academy and into the homes and neighborhoods of America."

I don't know that you would necessarily call it poetry, but there is something genuinely expressive, often even lyrical, about the way these words get strung together, sometimes with meaning, sometimes with a more random, absurdist narrative. Words drift together, then vanish in that improvisational flight that is so often the genesis of real thought and exchange. Nor does the alliance of words occur in any particularly linear progression; rather, the words float and hover around, above, and below one another until some connection between them is made. Or not. Things are half stated, then retracted. And while most of us are surely capable of making assured statements and confident commands, I remain certain that human communication, for the most part, is closer to this kind of fugitive, tentative endeavor.

The word tiles proved so popular over the years that in 1997 Workman published a collection of populist poetry, *The Magnetic Poetry Book of Poetry*, with a preface by the then U.S. poet laureate, Robert Pinsky, who wrote, "The word poet is based on the Greek word for 'maker,' which suggests that the artist in us is deeply related to the tinkerer, the gadget rigger who feels the urge to pile one stone upon another....Poetry extends that restless making-instinct into language." Even a reviewer with the rigorous intellect and commanding literary reputation of Sven Birkerts weighed in, suggesting in the *Atlantic Monthly*, "Not everyone can be a poet, but let's give out the magnetic words and cover the hard steely surfaces of the world with messages, charms, and barbaric yawps."

Certainly that has happened in my kitchen. There is a series of words on my refrigerator – "summer," "glad," "ice." But the composition does not stop with the word tiles from the little box. Recently I received refrigerator magnets illustrated with small Japanese landscapes – a branch of plum blossoms, a mountain path, a small fishing boat. Nearby, a magnet in the shape of a parrot head hovers. To the right, below, is a magnet with the name of our electrician. And drifting somewhere below them all is a decal of Allen Iverson, who is dribbling a basketball around the magnet of a bamboo bridge in snow. One day, the word "summer" is paired with "ice." The next day, "rain" has become "summer." Another time, the basketball player is dodging "glad."

There is no real connection between any of these words and images, yet their proximity of the surface of the refrigerator has forced them into an improvisational collegiality. And somehow, the way the small fragments of information and imagery drift in and out of one another's orbit day after day seems very real, communicating something essential about the facts of our family. What stands in my kitchen, I know, is a genuine infopliance. F. Scott Fitzgerald famously said, "The test of a first rate intelligence is the ability to hold two opposed ideas in the mind at the same time, and still retain the ability to function," and it is impossible not to think that while our refrigerator isn't smart in the current technological meaning, certainly it meets Fitzgerald's standard of intellect.

Decades later, Fitzgerald's elegant proposition has been streamlined, minimized, reduced to "multitasking," and there is yet another improvisational function served by many refrigerators. In numerous kitchens I have been in, the refrigerator, and the freezer in particular, serves as a kind of lock box, a safe, a storage unit. Perhaps partly acknowledging this practice, the Swedish manufacturer Electrolux AB designed a prototype in the late nineties for a refrigerator called the "Lighthouse." The appliance didn't have screens, but windows, and functioned as a kind of illuminated table, with the outlines of the objects stored inside appearing softly in the frosted glass sides. The thirty-six-inch-high (standard counter height) unit also came equipped with softly illuminated drawers and cabinets. Robin Edman, one of the designers working on it, says simply, "I tried to look at the object from a different point of view. Some people want to display their food. It makes them want to cook."

One reason the Lighthouse never went from prototype to production may be that it didn't go quite far enough; after all, it is not just food that people store in their refrigerators. For years, Luc kept a small plastic container with our dog's fur in it. "It's an experiment," he would tell me, but an experiment for what? I have known people who kept their jewelry in the freezer, the wing of a cardinal, a dead bat, the deed to the house, a will, all kinds of treasured items. That this appliance may also serve as a crypt or a tomb for treasured objects is only logical. Its temperature, its tight seal, and the pure incongruity of putting cherished items where perishable food is kept make it an ideal storage bin. That a salient characteristic of a tomb is silence underscores why such appliances should have the capacity for quiet as well as talk.

But when you think about it, it only makes sense that people want to put everything in their refrigerators, from dead bats and sapphires to anything else that matters to them. Preservation is at the heart of this appliance; it is all about *saving* things. Small wonder that in her grief at her grandmother's death, Sakurai found comfort curled up next to the refrigerator; its constant and comforting hum was the sound of small things being preserved.

And I wonder if this is why words and refrigerators are so innately connected, and why we are so compelled to give this appliance a voice of its own. Words, too, are agents of preservation; through them we preserve ourselves and the facts of our lives. We know this. It is why our words collect and gather there, whether they are the Internet-enabled commands of Screenfridge or the more tentative fragments of a magnetic word tile. It is why our notes and schedules and appointment cards and calendars all end up here. It is the reason Audrey failed, and why we want our refrigerators to be smart. It is also why NASA, in its efforts to re-create the comforts of home in outer space, might consider equipping its RFRs not with whimsically embossed food trays, but with ordinary refrigerator magnets.

Looking at the little magnetic word tiles, I realize that what I value most in them is their constant reminder that words stick. There is something about their very materiality that commands respect. Their most resonant message that a word is a magnet has a certain irrevocable power; here is physical affirmation that words fasten themselves to things. This is a wonderful thing to think of every time you reach for the milk. And surely any appliance that points it out countless times each day is very, very smart.

The SNOWBOARD

My father taught me how to ski when I was eight. We had recently moved to upstate New York, and determined to instill in my sister and me the same love he had for the sport, he bought us each a pair of blue wood skis. If they had a name, I don't recall what it was, because it was a time when sports equipment had not yet learned to say its name. I know my father's skis had no name. He loved skiing and for the better part of his adult life owned a single pair of long wood skis. These skis were not even painted a color – they were simply a varnished blond wood that had aged and darkened over the years to a deep honey. He would use these skis when he took his long cross-country runs up in the woods behind our house and at the local ski area, an hour's drive away.

Finally, when he was in his fifties, he broke down and bought a pair of sleek black metal skis that were relatively new on the market. They were called Heads, of course, and were the product of decades of research by Howard Head, who in 1947 had found a way to fabricate skis using strips of laminated aluminum. With sharper edges than wooden skis, they gave skiers more control and enabled them to cut sharper, more precise turns; easier to maneuver, the new revolutionary skis also made the sport more accessible to beginners. And there was something about the name of the ski, written in austere typography across its tip, that seemed to reflect the skis' superiority; while simply named after its inventor, this brand also invariably suggested that there was an intelligence, possibly even a cerebral quality, to this technically advanced piece of equipment.

Years later, I came across a postcard written by Anne Morrow Lindbergh. "Skiing is the intellectual's sport," she had noted, and I instantly thought of my father on those black metal skis carving their way down a snowy mountain in a direct, clear, linear path, a precise image for what Lindbergh meant. My father skied with the same elegance whether he was in the back woods or on a steep, groomed mountain trail. Even after he had grown accustomed to the Head skis, it never took much for him to mutter something about the frivolity of using two sets of skis for the practice of a single sport.

It's different for his grandsons. Even when they were young, seven or eight, nearly everything they wore had a message. All their clothes had an opinion – the logos on their baseball caps, the obscure witticisms on their T-shirts, the patterns on the soles of their shoes. The graphic, and very often narrative, content of their clothing was constant. It was no surprise, then, that a remark that they would commonly make in response to a parent's observation about news of the day, appropriate attire, or just about anything was often, "Too much information, Mom." It was a flippant way of saying they'd heard enough. And though funny at first, it later seemed a little sad and cryptic that kids under the age of ten would utter such a thing. Because it seemed to me that there can never be too much information for kids who aren't even ten years old; as the young citizens of something called the information age, they should be looking for as much as they can get.

But look at snowboards, which often seem to take information overload to a new extreme where it seems not only resolved, but logical, poetic even; a place where input overload makes its own sense. For graphic designers, the snowboard appears to be what the record album or book jacket once was, its imagery dense, varied, unlimited: Pop graphics, porn stars, graffiti, human anatomy, Gothic typefaces, tropical floral designs, urban skylines, Asian calligraphy, and pretty much anything else can serve as board graphics. Snowboards can convey more spiritual expression as well. Consider a series of boards designed by the studio Modern Dog Graphics for K2 in which a ray representing energy vortexes is encircled by a series of rings that refer to energy centers.

Before his sudden death in 1999, graphic designer P. Scott Makela developed graphics for an ad campaign for Rossignol that

were typified by typographic experimentation, energy, and technical skill. "God is close to ya," read one piece, and another, "Vertigo is fun," in a print campaign that combined oversize, expressive typography with crisp, fluid photography and dramatic imagery. More recently, Geoff McFetridge, a graphic designer in Los Angeles, has created hybrid graphics for snowboards that include floral patterning, camouflage, and urban landscapes. Of the graphic extremes snowboards so often reach, he says, "It all started out as a rebellion against the conservatism of skiing and all those stripes."

That sense of animation is hardly limited to the board's graphics. The snowboard itself is a product of constant, animated cross-exchange between products. Its earliest incarnation was the 1965 snurfer invented by Sherman Poppen, a chemical engineer who tied two skis together and attached a rope to the single ski, so that his young daughter could keep her balance as she maneuvered her way down snowy hills. Not long afterward, the Snurfer was licensed, but more as a child's toy than as a legitimate piece of sports equipment. In 1972, Dimitrije Milovich produced the Winterstick, a hybrid that brought the function of skis to a board resembling a surfboard. Five years later, Jake Burton, remembering the Snurfer he had used as a boy, laminated hardwood boards together and attached a binding, which made control feasible, and the sport as we know it today was born.

Since then, the cross-pollination between skis, surfboards, skateboards, in-line skates, wake boards, luge boards, snow blades, and snow skates has only continued. And while contemporary snowboards use the plastic bottoms, metal edges, camber, and sidecut of skis, they reflect a different sport and sensibility. In his 1998 essay "Notes on an Infinity of Sports Cultures," writer and designer Steven Skov Holt observes that by the nineties, "board sports [had] been merging, cross-referencing, and playing off one another in both their specific details and in their general culture.

Design [had] become a fluid medium of exchange, a process of communication between the different board factions, a way in which color, pattern, form, and detail can be used to signify allegiances and wordless understandings about what's hip and what isn't."

In sports design, this process of constant animated exchange is often called "technology transfer." But a better term for it is probably "morphing," a concise word for the inclination of sports equipment to beg, borrow, and steal from one another – the way roller skates became in-line skates using parts appropriated from skateboards; the way all-terrain in-line skates with oversize wheels were a hybrid of the skate and mountain bike; the way snow blades changed the shape of skis by considering the moves of snowboarders; the way street luge boards morphed from skateboards and luge sleds; or the way air boards (used by parachuters to carve air before opening their chutes) appropriated the shape of skateboards. And then there are such novelties as a snowboard that actually becomes a ski. The Voile Split Decision Twin Tip is a snowboard with removable steel plates that can either function to bind the two pieces of the board together to make a snowboard or, when released into a different configuration, serve as the ski bindings. While surely a niche, such hybridization remains a consistent theme in the sport.

Consider the snowblade, designed in the mid-nineties. Though clearly related to the traditional ski, it has a more parabolic shape, flipped up lip and tail, and shorter length, all evolved from the snowboard. Designed for halfpipes and jumps, it has been described as "a ski on amphetamines." Its designer, Vlad Zinovieff, said of it, "The snowboard culture has shown the industry a new attitude towards gliding. It's more of a culture and less of a discipline than skiing. With its sensation of carving and the same sense of balance you use in skateboarding and surfing, snowboarding became to skiing what punk was to rock 'n' roll. Snowblades have some of the same culture....There is a greater freedom of movement. A lot of people think that the snowblade is to snowboarding what in-line skating was to skateboarding. Myself, I just like to think that all these boundaries between different sports are disappearing."

A similar dissolution of boundaries was of interest to Tuukka Kaila, a Finnish skateboarder and photographer whose images combine the recognizable marketing graphics of known brands like Nike and Sony with familiar folk-art imagery and Soviet icons. His essential medium is convergence. In a statement accompanying a show of his work at Kiasma, a museum and cultural center in Helsinki, he wrote, "I've been interested in Soviet Russian things for a long time, and how to combine that with all-American skateboarding, the very brand-aware world of young people; it's not that much about a conflict, it's about merging and bringing things together."

While fusion may occur all across the cultural landscape, it is does so rarely with such exuberance and raw energy. Power yoga, Asian beef tacos, and hip-hop reggae are all evidence of cultural cross-pollination, but there is something about the very physicality of sports that gives this kind of fusion a different energy; somehow, a phone that functions as a camera or a camera that morphs into a gun doesn't have the same appeal of a piece of wood or fiberglass that can be reshaped to surf not a wave but a mountain. Board sports are the sports equivalent of attention deficit disorder: nothing stays the same for long; change is constant. Or you could just say it's about transformation; as surfing has moved from the ocean to the mountains to the Web, the inherent message is that not only are shape, purpose, and movement fluid, but so, too, is identity.

Holt wrote, "Given the barrage of information that flows over, around, and through us every day, we don't have the benefit of living in a time of static clarity. But ours is a time of distinct opportunity, and the design of sports equipment is capable of addressing these new benchmarks of complexity, confusion, and contradiction, in the end offering its own new form of clarity." Although referring to contemporary sports equipment in general, the comment seems especially relevant to snowboards, which seem somehow designed to navigate their way not only through rails and halfpipes, but through the dense information overload. The daughter of a friend of mine puts in another way. Of snowboarding she says, "It's almost Zen for me. It's nothing like skiing. Snowboarding is all about keeping your balance, your focus."

If snowboarding is about transformation, its documentation captures this. But while the camcorders used to document the sport use up-to-the-minute digital technology, their style of photography, oddly enough, seems to look to the stop-action series photography of Eadweard Muybridge, the self-taught, nineteenth-century photographer whose human and animal locomotion sequence studies seemed to arrest time. Muybridge's exposures dissected motion into visual fragments, and in their sequential records of image and movement, offered us mesmerizing stills of kinetic energy. He used multiple cameras, each of their shutters outfitted with a wire, which was tripped by the moving figure. In such a rudimentary fashion he recorded how a horse, a tree sloth, a stork, a cat with an injured leg, a man, and a woman might move forward. Aside from presaging motion pictures, Muybridge's photos suggested that movement could be visually analyzed and understood.

While the color images of snowboarders in my sons' magazines are many generations removed from Muybridge's trotting horses, the former might include fifteen or twenty shots of him (or her) twisting, turning, gyrating, spinning, bending in the air before making contact again with the mountain. And it makes sense that nearly 150 years after Muybridge's studies became known, snowboarders find a similar appeal in sequential imagery. I am told that the multiple views are made to clarify the moves and tricks, all of which often happen too quickly to be fully grasped. While the sequential pictorial narrative seems quaintly antique, its documentation is strictly contemporary: what it may capture most successfully is not the nuances or measurements of movement, but the sense of performance. The filming of snowboarding is often nearly as important as the snowboarding itself, and the aspect of exhibition, of pure theater has become a byproduct of many contemporary sports.

As I consider the multiplicity of images, the visual overload, the mountain as theater, I cannot help but think of my father traversing the mountain on his black skis. The picture has a clarity, a directness, a sense of purpose that all seem to reside in some parallel universe to the shredding, nosegrabs, backsides, and rotations practiced by my sons. Snowboarding is not what you would call "the sport of intellectuals," but it represents a different kind of thinking, one that has everything to do with how we accommodate complexity and contradiction. The streamlined single-mindedness we once so valued has been replaced by something else, and we seem to put a higher premium today on the ability to process multiple ideas quickly, to make connections, to process information in a lateral rather than linear way.

Years ago, I saw a snowboard that had appropriated the big blue, red, and yellow polka dots of the Wonder Bread packaging. It was about as far as you could get from my father's stark black metal skis. And it was an object of sublime logic. A product that cheerfully resonates with artificiality, Wonder Bread conveys a kind of retro optimism and has become a jolly little domestic icon. Who wouldn't want to surf down a mountain on this? I remain certain that the allure of snowboarding lies in its fluidity, its innate message about constant, irrevocable change. The greatest wonder may simply be in believing that one thing leads to another; that what could happen on ocean waves could be replicated on urban concrete and then on a snow-packed mountain is probably one of the more creative ideas offered by modern culture in recent years. It suggests that possibilities come in multiples – which is probably what kids who so often have too much information in their lives must have, must believe in.

My father died several years before his grandsons were born, but from time to time I imagine being given the opportunity to introduce them. The meeting always takes place on a snowy mountain. I picture him on his elegant black skis with their simple typography spelling "Head," shaking the hands of these boys on their Salomon boards, with logos, decals, type, bursts of color splashed across the surface in the haphazard constellation of imagery that inevitably seems to embellish their lives. He has been traversing the mountain, they have been shredding it. In my dream of this impossible meeting, their skis and snowboards have somehow become the accessories that have allowed them all to surf this greater distance.

The snowboard

The CAMERA

The industrial designer Henry Dreyfuss was often characterized by the conservative brown suits he wore, his innate dignity and businesslike approach to design, and his thoughtful and deliberate approach to the shape of things, be it a light switch or a locomotive. Yet he must have appreciated excess when he saw it. When he designed the Swinger camera in 1965, he recognized the sensibility of a market as far from his own as he could get. Dreyfuss's conviction that machines must be designed to fit people was expressed nowhere with greater clarity than in this camera meant to appeal to teenagers, who in that decade leaned to more extravagant behavior.

Modifying the Polaroid technology then available to a more high-end adult market, the Swinger was inexpensive ($19.95), easy to use, and fairly durable. In his book, *Henry Dreyfuss: Industrial Designer,* Russell Flinchum calls the camera "a hallmark of the 1960s passion for fun, plastic, and immediate gratification" and recalls that it was marketed with ads featuring Ali McGraw in a bikini, inviting young photographers to "meet the Swinger." The elegant leather and metal of the camera's predecessor in Polaroid history was replaced by white and black plastic suggestive of Pop art, and a solid body took the place of collapsible bellows. In not requiring a lab to process the pictures, the Swinger gave its teenage users a sense of illicit authority; it decisively transferred picture taking and making from the realm of science to the realm of pure pleasure. There was a tactile element to the camera as well; it had a physical presence. Your hands got sticky applying the coating to the film, which somehow gave you a sense of importance in the process.

The very idea of meeting the Swinger deviated from fundamental precepts of serious photography. In the early days of the craft, the large-format camera invariably established a formal presence and mediated a distance between photographer and subject, but as photography grew as an art form most of its practitioners became more interested in reducing that distance so their images

might be perceived as more candid or more real. The Swinger had no such conceit; social intrusiveness was at the very core of its being, and reveling openly in the inevitable interaction between user and subject, it was happy to serve as a small accomplice in one's social life.

I remember this because I had a Swinger. While it was not particularly small or compact or easy to carry around, if you happened to be a teenager then, you somehow found a way to take it with you. It came with a loop so the camera could dangle on your wrist, and as awkward as that might be, it was worth the effort because having that camera reminded you that the moment could be captured forever; experience could be stopped, visually examined, reflected upon – all practices that seem to come naturally to adolescent girls. Given the tenor of the times, that kind of occasional pause for reflection was not a bad idea.

A casual little box, a home lab for memory, the Swinger produced small, square images that documented the moment, though that moment was never quite as you had imagined it. The camera's technical capabilities were not equal to the mood it created; it was the promise of the technology rather than the technology itself that mattered. If the photographs could be produced instantaneously, it hardly mattered that the images tended to be flat and their gradations from dark to light had little subtlety. I have an old photograph of my brother diving off a board into a pool. The sunlight, the water, the board, the concrete all seem to reflect light off one another, and the afternoon has been reduced to a series of planes of light and several blurred images. But no matter. The Swinger's cheery spontaneity, its do-it-yourself technology, its portability, and its absolute dedication to the fun going on forever all rendered the camera an inevitable icon of the sixties.

I can't help but think of the Swinger when I see my sons and their friends with their digital cameras. Their Sony Digital Handycam is a video camera they use when skating, snowboarding, snowblading, tailsliding. Though configured differently than my old camera, it is roughly the same size, and it, too, is an active participant in the social dynamic of a group of teenagers. But any similarity with the Swinger ends there. The crucial difference is not so much that chemical emulsions have been replaced by electronic sensors, but that with those sensors comes a seemingly endless series of choices: images can not only be produced instantly, but they can also be amended instantly. From the screen that allows users to view images before taking the shot to the delete button they can hit to eliminate shots, the documentation of experience offers unlimited visual possibilities.

There is nothing new about the human impulse to edit photography; it's probably as close as any of us will ever come to rewriting the past. For years in the middle of the last century, for example, British newspapers routinely airbrushed out the cigarettes, wine glasses, ashtrays, and any other evidence of presumed dissipation in society photos of the elite; altering reality was simply another luxury available to aristocrats. Digital cameras and all the software that comes with them have simply made the process more democratic and efficient. Microsoft's Smart Erase program can eliminate undesirable components in a photograph – the cursor outlines the element, erases it, then fills in the empty background with the existing background. In the ad I saw for the program, a background figure wearing a red jacket at a ski resort had been made to vanish. It's no longer about ashtrays and wine glasses. People, furniture, and cars can all be instantaneously manipulated out of the frame. In architectural photographs, telephone poles, trucks, and billboards can be visually dismissed. And that staple of family albums – the photograph in which the divorced spouse or alienated sibling has been torn away – may be a thing of the past. The torn paper that is the vestige of these severed ties is gone. Absence is seamless.

Which, it seems to me, is not unlike the way we have come to view memory itself. That memory edits itself is something we have come to accept as scientific fact; and that our imaginations are already hardwired with Smart Erase is part of conventional psychoanalytic theory. Criminologists have told us for years that crime victims are unreliable witnesses, that eyewitness memories tend to be deeply flawed, and that memory is fragile, fluid, variable. During an assault or theft, fear can distort and deform memory. But elsewhere in life, too, desire, envy, excitement, sorrow, or any number of other common emotions are all forces that can bend and manipulate memory.

Smart Erase is the least of it; just as our minds obliterate past events, so can they create them. Human memory is open to suggestion. We have assigned the name "false memory syndrome" to what happens when we construct images and events based upon often the most subtle suggestions; and "memory transference" is the term used to describe what occurs over time when our own biases and beliefs superimpose visual images upon known people and events. One face is substituted for another, an afternoon becomes a morning. The image of the blue scarf lying in the drawer is stronger than the recollection of taking it out of the drawer. How we remember the events of our lives is a novelistic process. Memory is open to suggestion and often conforms to what we know and what we want to know more than to any actual sequence of events. Its distortions seem an inevitable product of how we encode experience. In the end, it may simply be human to be an unreliable witness.

If we accept the fact that memory is open to revision, is it any surprise that we have devised digitized tools that can be agents of this change? Is it any wonder that my sons' Handycam can effortlessly crop, edit, retouch whatever it records? The images they film can be expanded vertically or horizontally; contrast can be emphasized to make them resemble animated cartoons; selected backgrounds can be dropped in; images can be made to appear in sepia or black and white. A sequence can be viewed at slow speed, at double speed, or frame by frame. Soft backgrounds can be created, faces gently spotlit, distant landscapes made to appear close. Whatever appears on the screen can be enlarged or diminished at will, its color and brightness adjusted.

With the use of something called a Memory Stick, whole scenes can be played back, recorded, or deleted, while with Memory Mix still shots can be superimposed on those that move and words or entire blocks of text overlaid on images; one image can work as a scrim against which another appears. And with Memory Overlap, moving pictures can be made to fade in over still images. The instruction manual to the Handycam offers pages of directions, too, for the "fader function," in which images can be made to fade at any speed or to any degree. When fading in, pictures can be programmed to change from black-and-white to color, while doing the reverse when fading out. They can fade in and out with subtlety or in concisely delineated frames, in a mosaic of pixels or in a pattern of scrambled dots. When desired, the fader function can be canceled entirely. All of which is to say the Digital Handycam is equipped with features that mirror the character of memory itself; here is a camera that recognizes that memory arrives in layers, shaded in its own colors, cast in its own light, moving forward and back at a speed of its own determining, shaped and sculpted by some elusive conspiracy of the conscious and unconscious mind.

Surely writers of modern times have concerned themselves with the obscure avenues of memory – imagine what Proust could have done with a Handycam. If the digital camera had been described to us thirty years ago, we might have concluded that Gabriel García Marquez had not investigated the shadow world of magical realism in fiction, had not spent a life in letters imagining how nostalgia "as always, had wiped away bad memories and magnified the good ones," and had instead gone to work for the Sony Corporation to design cameras. But he didn't, of course, nor as far as I know did neurologists or criminologists or behavioral psychologists who have studied how the mind restores and revisits data contribute to the development of this camera. Nor were Sony designers and technicians motivated by the knowledge that when memories are visited time and again, their information can be easily changed and that they often reemerge slightly altered. The Handycam is, simply, a product of its time.

I remember the pictures I used to take with the Swinger – my mother standing in her garden, my sister on the kitchen telephone, my brother in the pool – in all of them, the brightness of the image, the texts that accompany them, the scale of the people within the landscape, and the fader function exist only in my mind. I return to the photograph of my brother in the swimming pool. It happened once, and the entire scene has been distorted by an excess of reflected light. And then I am shown an image from Noel's film. He is skating off a ledge on the banks under the Brooklyn Bridge. He does it in the film repeatedly. Sometimes the scene plays out quickly, at other times it is slowed. The last time he does it, the bridge and the banks fade. He does this against a soundtrack, an old Rolling Stones song that, strangely enough, I realize, could have been playing the day my brother took his swim. With all the embellishment of my son's film – its fading, its mixes, its overlaps, its manipulation of time – it is hard not to admit that the Handycam is an instrument that captures the elusive process of memory itself; the most modern thing about digital photography is that it reflects our knowledge about human memory.

That a Handycam has become a common partner in experience, an efficient accessory for reflective thinking is suggested in various recent films. In Sam Mendes's *American Beauty*

(1999), Ricky, the boy next door, records the minutiae of suburban life with his camcorder in a role that blurs the distinctions between voyeurism and documentation; and in Jim Sheridan's *In America* (2002) an eleven-year-old girl views immigrant life in Hell's Kitchen through the lens of her digital video camera. It is so common an accessory that more rudimentary cameras are sometimes valued for just that primitive quality. In the late nineties, Polaroid introduced the i-Zone, an inexpensive novelty camera that produced instant, tiny square images the size of postage stamps. And like stamps, these came with an adhesive backing so the images could be used as stickers; the notion that taking pictures could be a tactile experience was a distant echo of the Swinger. My sons used the i-Zones for a matter of weeks. They have also used disposable cameras and underwater cameras and cameras that allow them to format photographs as squares or in panoramic layouts. They know that there are countless ways to document and shape human experience. They are familiar with the idea that memory is a creative process.

So it should have come as no surprise when they recently asked for a "real camera," one, they meant, that uses traditional film for still photographs. There was an undeniable charm to the way they attached the idea of what was real to tradition. And I wonder if this is the appeal they find in the still shots of a traditional camera, which produce documents in a surer sense of the word than digital images, images that cannot be so simply altered, revised, faded, or superimposed. Perhaps it was the uncertainty of Memory Mix, or its implicit suggestion that past events are malleable, but my sons have become ready to entertain the quaint and outdated but clearly reassuring notion that reality is more fixed.

In her book *Regarding the Pain of Others*, Susan Sontag wrote, "Perhaps too much value is assigned to memory, not enough to thinking. Remembering is an ethical act, has ethical value in and of itself." She was discussing war photography and how we respond to images of pain and affliction in others, but her statement has broader relevance in this age of multiple images and imagery, and I imagine the possibility of these words being included in the instruction manual for the Handycam. It's useful advice, after all. But where in the manual would her words be the most meaningful? Possibly in the paragraph that tells us how to swap a still image with a moving image, or how to enlarge or reduce the size of images. Or possibly anywhere at all in its 210 pages.

In their ability to edit, revise, and otherwise visually reconsider and reconstruct the events of people's lives, the digital cameras my sons use may be precise instruments for recording human experience. The skier in the red jacket has vanished. The blue scarf is no longer in the drawer. The expression on their mother's face has faded beyond recognition. As these images are every bit as fleeting as memory itself, they cannot help but reflect the fugitive character of memory. While the Handycam is outfitted with a delete button, fader function, and Memory Mix, what it shows most of all is that memory is about what we choose to forget as much as it is about what we decide to remember – which, after all, is the way we find meaning in experience. Surely that makes it an ethical as well as an aesthetic process. Because what may matter most in the end is how we make those choices – what we give an added brightness to and what we choose to fade.

And I envision the spiritual partnership between Dreyfuss, working with his conviction that machines must fit people, and Sontag, driven by her own belief that human memory is a necessarily responsible process. In the archives of imagined collaborations, I remain certain theirs might have produced the Handycam of our dreams.

the MEDICINE CABINET

A woman I know who is well into her eighties told me recently of her health regimen. Although she takes medication for arthritis, it is her forty minutes of vigorous exercise in the morning and her shot of single malt in the evening that she credits with her good health. And she has the full support of her doctor. "He doesn't want to prescribe a lot of pills for me that I don't need," she said. "He also told me that he often loses the trust of patients my age because he doesn't like to prescribe a lot of unnecessary drugs."

I thought of her recently when I saw a bathroom renovation the high point of which was an enormous, sculptural medicine cabinet. Costing over $3,000, it took its decorative cues from luxurious bathing fixtures of the thirties and forties; yet for all its antique detailing, nickel plating, beveled glass, and piano hinges, resonant with details from lavish baths of old English country houses, this cabinet spoke clearly to its own time. Its oversize dimensions and generous capacity to accommodate an abundance of medications were strictly contemporary. I have seen photographs in magazines of floor-to-ceiling medicine cabinets stocked with every manner of cosmetic apparatus and have read about his-and-hers medicine cabinets that seem to suggest that well-being necessarily observes a gender bias. An even more likely scenario is the medicine cabinet positioned above the double sink and running its entire length, thereby doubling and sometimes tripling the space of the conventional cabinet. That these cabinets have become taller, wider, deeper in recent years only makes sense – at a time when health and healing resonate with anxiety for so many people, it follows that the medicine cabinet has grown in size.

Filled with the real and imagined accessories for well-being, there is something at once sinister and poignant about the medicine cabinet. Its restorative powers evolve from a convergence of the pharmaceutical and the cosmetic, and outfitted with a mirror for the more elusive process of personal reflection, it becomes a

toolbox for the self. Its contents, at once random and selective, invariably reveal our weaknesses and fears along with our hopes and desires – painkillers, Prozac, perfume, and whatever other assorted accessories we rely on to piece together what we may consider good health. Like women's pocketbooks or a car's glove compartment, the medicine cabinet has become a kind of cultural grab bag, its interior brimming with intimate information, revealing details both personal and cultural. It is no surprise, then, that gazing into your date's medicine cabinet has become part of the contemporary dating ritual to the point that its ethics were discussed on *Seinfeld* in an exchange between Kramer and Elaine: "I always open medicine cabinets." "I trust people not to do that." "Big mistake."

As an artifact with a potency of its own, it makes sense that artists and writers have found the medicine cabinet an image resonant with ideas of beauty, fear, health. In J. D. Salinger's classic novel *Franny and Zooey*, the account of a family's limited progress with psychoanalysis and religion is preceded by a list of more basic accessories for what people may turn to to make themselves well: the full-page inventory of the family medicine cabinet itemizes everything from iodine, Mercurochrome, two Gillette razors, and six bars of castile soap to a torn snapshot of a cat, seashells, a school ring, and ticket stubs for a musical comedy. More recently, the architectural team of Diller + Scofidio found the medicine cabinet a domestic object that reflects our relationship with pleasure and pain. The Pleasure/Pain Medicine Cabinet they assembled in 1992 was an arrangement of objects that "reiterate that one person's pain may be another person's pleasure." Still, the things placed inside it were oddly antique – an old sea sponge, ace bandages, apothecary jars, all medicinal artifacts laden with nostalgia, the accessories of well-being from another time. Yet such a selection may have its own logic; nostalgia offers its own comforts.

Such nostalgia often seems a staple in contemporary bath renovations. While these rooms are now replete with the latest in spa technology – showers with myriad adjustable jets and sliding body sprays and tubs with built-in water heaters – they also tend to be outfitted with antique medicine cabinets. Perhaps because what is inside catalogues our fears so precisely, it may only be logical that the shape of the cabinet itself is reassuringly familiar; whether antique French Provincial, early American country pine, or embellished Victorian, each offers its own vintage brand of comfort. A simple and straightforward Mission-style cabinet made of oak cannot help but suggest that well-being, like furniture, can be crafted with a combination of elegance, practicality, and common sense. I have also seen retrofitted Chinese medicine cabinets equipped with brass fittings and an assortment of tiny drawers in an effort to convey the mysterious healing legacy of Chinese herbal medicine, though they may, in fact, contain an archive of strictly Western pharmaceuticals. But so it is in an age of pluralism. Design is often a way to make the disconnect charming, bearable.

Notwithstanding the implied comforts of French Provincial or vintage English country style medicine cabinets, a handful of designers and manufacturers recognize the cabinet as something that accommodates a more contemporary sensibility. Thomas Eriksson is a young Swedish designer who has devised a cabinet in the shape of a red cross for Cappellini. Despite its simple, modern geometric form, the brightly lacquered steel-plate cabinet is a stark icon loaded with more visceral associations, suggesting both imminent catastrophe and the message that help is at hand. Elsewhere, one finds more prosaic attachments and accessories that are purely of the moment. Robern, a company in Pennsylvania that manufactures high-end medicine cabinets, offers such features as interior electrical outlets, door defoggers, safety lock boxes with which to secure prescription drugs, mirrored interiors, and customized lighting.

Those consumers whose preoccupations with security follow them into the bathroom can attach a small, battery-operated electronic sensor with a motion detector to the cabinet to monitor when it has been opened. The research labs at Accenture have developed a prototype for the "smart" medicine cabinet that includes a display screen and miniature camera able to recognize the facial features of the person looking at it. The software can be programmed to recognize the label of the prescription bottle to identify the medication, then to post the correct dosage on the screen, along with its compatability with other medications. The cabinet is also capable of reordering prescriptions and issuing warnings should someone reach for the wrong prescription.

Most medicine cabinets, of course, speak a simpler, more traditional language. An exhibition called *Shelf Life* held at London's Two10 Gallery in 2001 studied the revelations of the medicine cabinet and the myths of home healthcare it contains. Catering to our voyeuristic impulse to peer into the private lives of strangers, it displayed fifteen medicine cabinets loaned by members of the general public, each accompanied by a personal statement about its contents. What critic Robert Hughes once called the "gentle fetishism" of Joseph Cornell's boxes could be applied to these as well. Cornell's boxes were filled with objects found in flea markets and junk shops, and the contents of the cabinets conveyed a similar random quality; while some things seemed to be selected with care and deliberation, others accrued more erratically.

A plain, wooden, two-shelf cabinet was accompanied by the statement, "There are things that shouldn't be here. The lighter fluid and Ventolin, a scalpel, some pills my brother's girlfriend was taking." There were wheeled carts, a plastic bucket kept on top of a bureau, a metal cabinet picked up at a flea market, and a three-door plastic cabinet whose crowded shelves mirrored the battle for bathroom time in a small house. An old-fashioned wooden first-aid box with tiger balm, bandages, and assorted ointments was accompanied by arch comments alluding to a presumed invulnerability: "We hardly ever use the box." Another lender noted, "I try to avoid using medications and to be calm about unexpected and strange phenomenon [*sic*] that suddenly attack and occupy one's body."

Yet for all their poignance, most cabinets hardly need such explanations. As storage spaces for the self, they are able to speak well enough on their own. And for all their varying views toward health and well-being, consistent in these overburdened cabinets is a sense of physical intimacy. Chemistry kits for personal health, they demand lavish and persistent scrutiny; we are advised to keep them clean, to keep prescriptions up to date, and to tend to their contents, which change as constantly as we do ourselves.

The devotion they elicit sometimes positions medicine cabinets in a universe parallel to that of the small Buddhist and Shinto shrines often found in Japanese homes, where small gifts, rice, fruit, and sweet cakes are put out to honor family ancestors. Cabinet and shrine share a proximity to and a respect for the unseen, although in the case of the former it has to do with a fear of and fascination with the microscopic world. Contagious diseases were once something we were simply aware of. Today, AIDS, SARS, cruise ship viruses, and other contagious diseases seem more threatening, and there is a preoccupation with things unseen, a cultural mania for obsessive cleanliness. An aversion to shaking hands or touching doorknobs is no longer considered eccentric but sensibly hygienic, and over one thousand antibacterial lotions, soaps, detergents, and other household products are available on the market. The medicine cabinet is no less efficient or compact than the shrine. A place that allows us to nurture our phobias, it is full of the same expectation and desire for safety, health, and happiness; but the articles inside it are collected to honor oneself, not ancestors or god. And like those small altars, it invites a ceremonial reverence.

Because it is, of course, the unknown and unexpected that we are talking about here, which may account for the growing dimensions of the medicine cabinet. As self-diagnosis and self-treatment become increasingly common means of getting medical care, this furnishing has more to contain today. The need for self-treatment has led to innovation in the design of all types of home medical accessories, from pill splitters to syringes. The Swedish studio Ergonomidesign, for example, devised a syringe that resembles a pen; because it was intended for use by patients, many of them kids, who are afraid of getting shots, the needle itself remains concealed in this benign familiar object. IDEO, the California-based design firm, regularly introduces innovative home healthcare equipment – HealthBuddy, for example, is a tabletop home monitor that asks patients routine questions about diet and medication, then sends the information to a service center that can be accessed by a doctor; and Body Gem, a calorimeter for home use. With a mask and mouthpiece, the unit measures oxygen consumption to determine the body's metabolism at rest and during exercise and can be used by people who need to track health and diet to assess their caloric needs.

At the same time, however, less reliable home labs and medical self-testing kits have found a growing market appeal. Despite apprehension from the medical community, people find such tests easy to use, inexpensive, and reassuringly private, and they rely upon them increasingly for both diagnosis and treatment. As people live longer and longer, home care for the elderly now includes ways for them to analyze and treat their own conditions, and kits can test for everything from such common ailments as the flu and urinary tract infections to HIV and Alzheimer's. (Though not without a certain intrigue, the scent-strip quiz used for the last has provoked genuine alarm in the medical community for its unreliability.) Blood pressure and cholesterol can be tracked as well. And although kits manufactured outside of the U.S. have not been tested or approved by the FDA and are illegal to market in this country, they are still easily available online. Their confidentiality and low cost have such appeal that their frequent unreliability seems but a minor quality flaw.

There is a world of difference, of course, between legitimate home healthcare equipment and the growing tendency we seem to have to take our health into our own hands with more questionable kits, home labs, and the largely unregulated industry of medicinal supplements. But so often, what we use and the way we use it may fall into a gray area that is somewhere between taking care of ourselves and endangering ourselves, and the medicine cabinet occupies that gray area. As the word "home" is employed increasingly as a prefix for "healthcare," as the costs of both healthcare and health insurance continue to rise, as the number of uninsured individuals continues to grow, and as adequate care continues to be out of reach for so many people, it only follows that the medicine cabinet has become a more substantial piece of furniture. No longer simply an accessory for benign self-reflection, it is furniture for a population disenfranchised from institutionalized healthcare, conveying self-reliance and independence from the diminishing comforts of a costly health industry.

Most of all, the medicine cabinet is a place of contradictions, its contents articulating a confusion about how we confront our frailties. It manages to accommodate pleasure and pain, beauty, well-being, nostalgia, up-to-the-minute technology. It expresses all the assurance and sufficiency that come from taking care of ourselves, from individual choice and responsibility for one's own health. But at the same time, the crowded shelves reflect apprehension and distrust for the pharmaceutical industry and its growing reputation for fitting diseases to the drugs in its research labs rather than the reverse; and the sense of abandonment at a time when established medicine seems to be at a greater remove for so many. A small tableau for self-examination, it is a place of mixed messages.

That sort of ambivalence was reflected in Damien Hirst's *Pharmacy*, an installation at London's Tate Gallery in 1992. Hirst lined the gallery walls with glass cabinets, a highly ordered grid of pharmaceutical supplies. The pill jars and medicine packages were arranged according to color and shape; in some cases, their placement on the shelves reflected the part of the body they were meant to heal. The composition was highly deliberate and ordered and served as a response to both the chaos that unexpected illness visits upon our lives and the sense of precision and efficiency we anticipate when we turn to medicine. Yet all the packages and bottles were empty. "I think art is a hell of a lot better for you than medicine, in the long run," the artist said in a statement accompanying the show. "You don't get a long list of side effects – or maybe you do."

That sense of vacancy seems thematic when contemporary artists turn to pharmaceutical imagery. The exhibition *Shelf Life* in London also included the work of artists. Joanna Walsh constructed a subtle and elusive assemblage of the glass shelves of a medicine cabinet, empty but for the rings, rust stains, pools of color left by jars, bottles, and tins, all the vestiges of the accessories of self-reflection and renewal. Like Hirst's empty bottles and jars, the vacant cabinet with all its remedial ghosts speaks as eloquently to our times as the oversize, vintage English cabinet stocked with every form of the cosmetic and clinical. Because this is what we all know: we can do everything for ourselves, and nothing.

the WHITE LAWN CHAIRS

My mother and father have been gone for years, but if I could meet them again, I would hope to find them at a cocktail party, in July, in the rose garden of our old house in upstate New York. It would be dusk and the flares would be lit. And we would sit in the white lawn chairs.

*The white lawn chairs.* These are the words that say everything about those lost moments of a family. They are more evocative and more precise than any others. They are about languor and repose, and the exchanges that sometimes occur when one's mind is at rest. Like the cocktail party itself, they are about saying absolutely nothing and absolutely everything and about not knowing, perhaps until a later time, which is which. And they are about the great beauty of such confusions. What they describe, too, is the smooth assembly of planes. With their straight backs and long, wide arms gleaming white, the lawn chairs stand like small, immovable monuments, champions in the obscure garden of family relations.

There is a logic to the fact that these chairs evoke family, because they were originally designed by a man so that his family would have a place to sit. While what we think of as Adirondack chairs have been adapted, revised, and redesigned over the years, the original prototype was the Westport Plank Chair, designed and constructed in 1902 or 1903 by Thomas Lee, a wealthy businessman who vacationed with his family in Westport Mountain Spring in the Adirondacks. Prior to that, the Adirondack chair had been a rustic construction of unfinished twigs and branches of indigenous woods, usually beech, ash, birch, or hickory, nailed together in a primitive assembly. The simple, improvisational quality of the chairs gave them their charm and identity, and they were marketed nationally, finding ready customers in the cult of naturalism that the industrial age had spawned. Durable, sturdy, and suggestive of a respect for unspoiled nature, the furniture embodied qualities that appealed to the class of wealthy urban industrialists when they migrated to their vast summer camps and cottages.

Lee's chair was slightly more formal. He constructed it in various proportions, testing the different versions on the members of his large family. But consistent in all of them was that each was made from a single wide plank, often as much as eighteen inches across, and each had broad arms. Wedged and nailed together, the planks created a form at once austere and voluptuous. The seats were deep and the backs deeply angled, and the slightly curved arms wide enough to hold a book or a glass. Lee offered the design for the chair to his friend Henry Bunnell, who was in need of income. Recognizing the value of both its simple construction and its suitability to the mountain camps, Bunnell took a patent out on the chair in 1904 (without telling Lee) and continued to produce it for the next twenty years. The chairs were made of hemlock and painted dark green or brown.

It was not long before the chairs turned up on the cure porches of the tuberculosis sanatoriums that had been built in the Adirondacks in the late nineteenth century. The sanatoriums offered an experimental and alternative treatment for the disease in the days before there was any known remedy. The cure, as it was known, included primitive inhalers, a strict diet, enforced social-ization with other patients, and a steady intake of cool mountain air. Accounts of patients' lives at these mountain sanatoriums resonate with displacement, the social ostracism of the ill, and enforced exile. Yet at the same time, the inevitable intimacy among the infirm, their tentative future, and the sense of pastoral remove also contributed to an aura of romantic sojourn.

The chairs in our garden were salvaged by my grand-mother from the porch of an old sanatorium in the Adirondacks. Multiple coats of white paint have long since covered the original brown or green. Yet just barely decipherable under these layers are

Bunnell's name and patent imprint, the letters apparent only in the faintest of braille, which can be better felt than seen. When I was growing up, my father pointed out to me repeatedly that if I leaned back in one of these chairs, rested my arms on its arms, and breathed deeply, I was positioned in such a way that I couldn't help but fill my lungs with air.

While Lee most certainly did not design these chairs for sanatorium patients, such was nonetheless the brilliant career of the chairs – they enhanced breathing! This was the common wisdom passed down from a time when no one gave a thought to "human factors engineering." Years later, when I learned about ergonomic design, I read that a "good chair should embrace you, supporting your whole body from head to foot," an edict that made me think not of well-cushioned, padded, office chairs that offer lumbar support, but of the white lawn chairs in the garden. These were chairs that induced a calm, and years later when I did yoga, trying to master the discipline of *ujjayi* breathing, I would often try to put myself back in those chairs, breathing in that garden.

The provenance of the chairs is almost as ghostly as the chairs themselves. Although there is certainly no documentation of any formal connection, I am certain that the lawn chairs are not such distant cousins of the Red/Blue chair designed by Gerrit Rietveld thirteen years later. That chair is an icon of De Stijl design, a movement founded in the Netherlands in 1917 that advocated a focus on basics – color, form, space, a pure reduction to the essentials. With its straight planks of bright color, black supports, and geometric composition, Rietveld's Red/Blue chair is a manifesto for minimalism, intended not simply to accommo-date the human form, but to express an entire philosophy of modernism. The furniture equivalent of a Mondrian painting, it reduced the notion of a chair to a series of planes and lines. Certainly, it is a graceful expression of those essentials but, in truth, it is minimal only if you think a study of line, space, and form is minimal.

The white Westport plank chair in our garden and Rietveld's Red/Blue chair share an angularity. Both also reflect a respect for the Arts and Crafts movement, which put a high value on simple construction in the age of industrialization. Rietveld also often had his own large family in mind, designing brightly painted toys for his children and furniture prototypes. While there is nothing to suggest that Rietveld ever traveled to upstate New York to meet Lee or Bunnell to discuss furniture design, the chairs they all built seem to be on familiar terms with one another; in the world of objects, as in life, things reside in parallel universes. Like the art and architecture of the modern movement that set out to simplify the way people lived, Rietveld's chair was meant to reconsider and reinvent the very idea of a chair; surely those similarly angled plank chairs in the Adirondack Mountains a continent away also addressed starting over, though in a very different way.

Rietveld once famously remarked that "sitting is a verb," suggesting that a modern chair constructed in modern times might offer not the nostalgic, indulgent comfort of a Victorian wingback, but awareness, thoughtfulness, mental acuity. It's a tall order for a chair. If sitting is truly a verb, I would be more inclined to think of the patients cloistered in treatment in the Adirondacks, contemplating the anguish of their illness, the liaisons formed in those remote treatment centers, their unknown futures, all while sitting still in their capacious chairs.

Decades later, the Red/Blue chair was reinterpreted by a landscape architect named Lester Collins for his Asian garden, called Innisfree, in New York's Hudson Valley. Slightly sturdier than their predecessor, constructed simply of weathered pine, and devoid of color, the Collins chairs reduce the Rietveld chair even further. They may be even more elemental than the De Stijl chairs, and certainly they bring their own composure to the contemplation of an Asian garden. They, too, are minimal, although it occurs to me that there is nothing remotely minimal about a good chair in a garden. Collins's weathered gray chairs could be the offspring of Rietveld's icon and Lee's chairs for his summer cottage. Rietveld once said, "To me, a work of art is a creation through which we joyfully share in experiencing a bit of reality," and while he was probably referring to the art and architecture of the early twentieth century, surely his words also apply to the design of a good garden chair.

Anything as basic as the Westport plank chair invites all sorts of interpretation and embellishment. One especially surprising version of the chair was recently installed at the recreation center and pool at our local town park. These examples have been modeled after later versions of the Adirondack chair, where the single wide plank has been replaced by several thinner planks of wood, which are both cheaper and easier to come by. This generation of chairs has been fabricated in a dark green plastic with a faux wood grain, a texture deviating from the smooth planks of the original. But so it often is when the form of things evolves or when something is reinterpreted; the inherent characteristics of the original are often referred to in labored and overly obvious ways.

That said, there is a different authenticity to the plastic Adirondack chairs. Despite their synthetic material – or in fact, *because* of it – these chairs are perfectly suited to their function, which in a sense makes them genuine siblings of the original. While their generous proportions are true to the original, the chairs are lightweight enough for little kids to drag around for different seating arrangements. The wide back of the plastic chairs is a good place to dry a towel, the wide arms a good shelf for your sunglasses, sunblock, magazine. In the clamor and exuberance of a hot August afternoon, there is no question that these chairs and the pool they sit by are about as far as you could ever get from the serene contemplation of an Asian garden, and farther still from Rietveld's austere views of how elemental form could generate new ways of living. Still, they are the coordinates of a summer day, the accessories to a total immersion in the outdoors, in the sun and near the water; and when you think about it, a congregation of these plastic chairs is an eloquent expression of Rietveld's idea of the joyful sharing in experiencing a bit of reality. Although that bit of reality is a public park in upstate New York, I have little doubt that they are also the distant partners of Collins's chairs at Innisfree.

That's the thing about these lawn chairs; they just go on repeating themselves, time and again, in one iteration after the other, familiar forms reinterpreted in new materials and colors, all of them as reflective of current times as the earlier chairs were of theirs. The painted wood may be replaced with recycled plastic. One edition I have seen is built from Peruvian mahogany, which its manufacturers must necessarily claim has been contracted from scrupulous timber management programs rather than from loggers who clearcut forests. The slats of wood may be striped, the arms curved, and in one recent version the classic fan-shaped back has been modified to a circular form, an Adirondack chair on steroids. Marine enamel paints in bright reds, lime green, sky blue may replace the austere white or dark green. Another model reproduced the traditional slats of the back but repositioned the seat on blocky legs, giving it a postmodern, eighties sensibility, as though Bunnell had teamed up with Ettore Sottsass to design a prototype for Target.

In the late seventies, the minimalist sculptor Scott Burton constructed his own version of the white lawn chairs, though his happened to be surfaced in pale yellow Formica. With squared-off arms and elongated, exaggerated rear supports, the chairs suggest cartoon characters of the original. I saw them once not in a garden or on a porch, but in an art gallery, where they asked the inevitable question, Why would a minimalist artist take on a form that was already as minimal as it might ever be? Ironically, Burton's chairs seem to be almost decorated, kitschy when compared to the original. Burton often called his furniture "pragmatic structures" so that such constricting labels as "art" or "design" might be avoided. Yet even that term seems too elaborate; surely it is less resonant that the words "the white lawn chairs."

Still, I remain certain that our lawn chairs from the porch of the sanatorium are the most basic of all. While design historians may have the conviction that the Rietveld chair is a study of basics, I am convinced that the administrators of the sanatorium were in fact the people who best understood the character of the Westport plank chair, because what can possibly be more basic to the function of a chair than being a place to breathe? The chairs in our garden are more voluptuous than De Stijl's or Collins's, and besides, they are white. Apart from drinking and breathing, these chairs also generated other, smaller rituals. Every spring we would repaint them. When my sister and I were kids, we were given twenty-five cents each to do the job. Years later, when my boyfriend and I decided to get married, it happened to be in the spring, and there was nothing better I could think to do than give the chairs a new coat of paint, the ritual another distant echo of Rietveld's conviction that a new chair could signal a new way of living.

This is how it is with the chairs, objects in which furniture and metaphor have become one hopeless and beautiful tangle. And they are there again this May, this same tangle, this same collaboration of stillness. It's a different garden. Where I have put them now, the lawn is shaded, and you don't often think to go sit there. Even though they are nothing but lawn chairs, they discourage casual use. In their profile of perfect repose a sense of authority resides. Oscar Wilde spoke of hoping to live up to his blue-and-white china. I aspire in the same way to sitting in the white lawn chairs at the far end of my garden. But they have come to make up their own formidable company. They are their own perfect assembly of three and make a good case for furniture going into exile. Possibly there is some irony in the fact that chairs designed for breathing should now appear as some spectral presence in the garden, glowing white monuments to people long gone. But this, too, I suppose, is how objects acquire meaning – they begin by signifying one thing, then start to represent the very opposite. They are filled with contradictions, at which point they become embedded in one's psyche.

At dusk, especially, their whiteness seems almost to glow like a sequence of bright flags. You could call them phantom furniture. And it is my great good fortune to have ghosts such as these in my garden.

Afterword: The Golden Slipcover

I grew up in a house where frivolity still had a good name. This was largely due to my mother, who like so many serious people understood the meaning of the word, and its place in life. She was adamant about being well read, articulate, and informed in political argument. But she was just as capable of acts such as goldleafing a chocolate mousse.

Frivolity, as I came to understand it in the kitchen and elsewhere, had a serious side and as often as not required discipline and labor. It demanded work. One Christmas, my mother spent five hours cracking open the shells of walnuts, emptying them of the meat, and filling them up again with pennies, tiny dolls, and bits of costume jewelry. Then she glued the halves together and put them, along with some unadulterated nuts, in an immense basket, where they were meant to astonish and delight small children. That there was a place in life for a sure sense of the frivolous was never questioned in our family.

My mother belonged to a generation of women more inclined, when depressed, to arm themselves with an extra spritz of Chanel than to discuss the origins of their unhappiness with an analyst. I would never question her logic, because my mother and her friends operated with their own kind of courage. Indeed, the fact that frivolity, pure nerve, and a sense of purpose could coincide so gracefully was made clearest to me many years later, when she faced a malignant brain tumor. Her three-and-a-half-year battle began with the partial paralysis of her right arm. As that part of her body was gradually stilled, so, too, was her speech.

The seizures she suffered as the tumor grew were a physical assault to her brain. One result of this assault was aphasia, a speech impairment that makes it difficult or impossible to fasten words to thoughts. The clinical definition of aphasia is "defect or loss of the power of expression by speech or writing or signs, or of comprehending spoken or written language, due to injury or disease of the brain centers."

When you have lost the ability to assign words to thoughts, there is not much point in talking. For my mother, giving up good talk was relinquishing a lifeline; so for months, she tried. Yet the clarity of thought and speech that had been integral to her life had been obliterated. Now, when she wanted a cup of tea, she would ask you for a hairbrush. If what she really wanted was a sweater, you might find yourself preparing her a salad. When I got married, on a blazing, sunny October day, she turned to me and said, "Darling, you know it's meant to be good luck when it rains at your funeral."

The strangeness of such a sentence was nothing next to her humiliation after she spoke it. The confusion my mother felt in trying to put simple sentences together outweighed the possible comfort she might find in her old friends. Except for family members, then, she stopped seeing people.

But one afternoon as she was resting in the library after lunch, she noticed that the slipcover on the chaise she was lying on had become frayed. The chair was covered in gold silk she had bought in Thailand some thirty years earlier, and it was worn out, torn, graying. I'll never forget what she did next. She got up, walked over to the telephone, and called an upholsterer.

In perfectly clear language, she made arrangements to have the chaise recovered.

You could call this an episode of clarity. Certainly it was that to her, because for a moment a golden slipcover had restored all the precision and lucidity of language that were dear to her; on account of it she had recovered her voice. It was a moment of clarity for me as well, because I understood then that frivolity is not necessarily trite or foolish or petty; rather it is about the way essential information often comes to us, unpredictably, through play. I also understood that there are times when frivolity can intersect precisely and perfectly with a sense of purpose. This can happen most effortlessly and most gracefully when a sense of purpose elsewhere in your life seems to be either absent or irrelevant.

Since that time – and probably in one way or another because of it – I have made it my work to write about design, about spoons and slipcovers, hats and houses. Sometimes the objects are called "artifacts of the physical world." I call them things, because so far as I understand it, design is about people and things. You could say that I write about design because I am fascinated by the relationships people forge with *things* and by the inevitability of how we engage in play with our material possessions. In my mother's case, she did it because the loss of speech made her relationships with other people unbearable. So she turned to the chaise and its golden slipcover, and for a moment on the telephone she was herself again.

People often think of design as having to do with style. But I think it has more to do with the mysterious compulsion we have to turn to things when we find it impossible, for one reason or another, to turn to people. What is it about objects that induces us to include them in our emotional life? I wonder, too, about the great comfort we seem to derive from the possession of certain objects. Most of all, I am astonished by the power we so often draw from frivolous accessories when we confront our fiercest battles.

Acknowledgments

For her years of insight, steadfast encouragement, and sound editorial counsel I am indebted to Susan Szenasy, editor in chief of *Metropolis* magazine. The continuing support of publisher Horace Havemeyer III has been no less important. Sharon Gallagher, Lori Waxman, and Melanie Archer at D.A.P. all provided essential guidance and direction when it was needed, and I thank them, along with my editor at Metropolis Books, Diana Murphy, whose assured editorial hand, patience, and humor were mainstays. I have been exceptionally fortunate in having George Skelcher's precise and evocative drawings accompany these essays. Meryl Pollen's design for the book established its own brand of clarity on the ideas here, for which I am grateful. A number of architects and designers have been generous over the years in sharing their thoughts with me; they include – but are in no way limited to – James Biber, Chris Capuozzo, Jennifer Carpenter, Robin Edman, Peter Girardi, Geoff McFetridge, Tom Newhouse, and John Scofield. Pat Beard, Anne Kreamer, and Mary Ruefle gave advice and support at timely moments. Lastly, my gratitude goes to Brian, Noel, and Luc.